Forest H. Belt's

Easi-Guide to SHORTWAVE LISTENING

HOWARD W. SAMS & CO., INC.
THE BOBBS-MERRILL CO., INC.
INDIANAPOLIS • KANSAS CITY • NEW YORK

FIRST EDITION

SECOND PRINTING—1975

International Standard Book Number: 0-672-20961-6
Library of Congress Catalog Card Number: 73-75084

Preface

"Tune in the world." Sound glamorous? It is. To a special group of people who spend leisure hours fiddling with the dials of a shortwave radio, nothing compares with the thrill of hearing a station from the other side of the globe.

Shortwave listening as a hobby has been around for many, many years. The ranks continue to expand. A real upsurge developed after the transistor brought better equipment at lower prices. Some shortwave receivers are still expensive, but you can also get pretty good equipment without spending an arm and a leg.

You can't always spot a shortwave listener. They come from all walks of life. Students, doctors, factory workers, farmers, lawyers, housewives, race drivers, airline pilots, secretaries, even unemployed school dropouts—all sorts listen to shortwave radio.

Your next-door neighbor might be an SWL. There is little outward sign. His basement could house a shortwave listening station with a half-dozen receivers; yet your only clue could be a slender thread of wire stretched between two trees in his backyard. A few listeners put up fancy outdoor antennas to pull in distant stations. But you can't know for sure what all that stuff is for unless you ask.

Not that shortwave listeners are ashamed of their hobby. They're not. An avid shortwave buff will talk your ear off about it. But he'll seldom bring up the subject himself. For one thing, many people can't understand the excitement he feels. Others associate shortwave radio with tv interference; that's ridiculous unless he runs a transmitter, and shortwave *listeners* don't.

Ask around. Your clergyman may listen to shortwave radio. Your insurance man could be really big on police and fire calls.

Shortwave listening is rewarding. You may be astonished at what there is to learn from the activity. You'll discover that a shortwave listener becomes vastly well informed about the world. He also knows what's going on in his own town.

This book leads you comprehensively through the world of shortwave. It could make you a casual listener, or a serious SWL. Whether you listen to news from around the world, music from some favorite country, or the police and fire service in your own vicinity, you'll find something of interest on the shortwave radio every minute of the day or night. The following pages tell how to get the most from shortwave listening. Some chapters explain equipment, others tell how to use it. You will learn what antennas do best. All in all, you'll discover what you should know to become a satisfied shortwave listener.

Several individuals and organizations contributed help and information for this book. Thanks go particularly to Gary Atkins, Steven Becker, Larry Belt, Edwin L. Cambron, Linda A. Cummings, Jay Hendrix, Donald J. Lange, Jim Lovell, Mary Mulligan, Jim Newman, Raymond G. Smith, J. B. Wathen III, Don Williams, American Radio Relay League, George Drake Associates, Consulate General of Chile in New York, French Embassy Press and Information Division in New York, Hallicrafters Company, Heath Company, Radio Shack, RCA, Regency, Trans World Airlines, U.S. Army, U.S. Information Agency, U.S. Navy, Western Electric Company, and Zenith.

Forest H. Belt

Contents

Chapter 1

A Hobby to Fascinate and Educate

The world is not very big anymore. Its size depends on your point of view. Here, for instance, is what our Apollo astronauts saw from 240,000 miles out in space. A view like this only stresses the fact that planning nowadays needs to be on a worldwide basis.

We must learn to think in terms of a global community. Being concerned about one country—or even one continent—just doesn't cover the situation. Like it or not, the inhabitants of planet Earth are living today in one big spherical community.

You can learn a little about the world and its doings from books. A typical schoolroom class in "social studies" exposes youngsters to a variety of information regarding the world community and its inhabitants. They learn who lives where, what languages they speak, maybe even their physical appearance. Seldom do classroom courses impart much understanding of the influences that shape individuals or groups in other parts of the globe.

Books, magazines, and newspapers help. If you buy and read what's available, you do acquire some ideas about people of other countries. You may even read analyses of *why* things happen there as they do. But books and magazines written and published in the United States present only our viewpoint. It's not like hearing what people of other countries say about themselves—or about us.

Imagine hearing the news of a country directly from the people it's happening to. They report world events with whatever bias exists in their country. How much more revealing this kind of "studying" is to a student of world affairs. (And remember, world affairs today are the affairs of our *total* community.)

The listener to shortwave radio gains this global viewpoint. Overseas broadcasts, heard with regularity, quickly acquaint the listener with differences in political systems. He hears many versions of what is important in a day's happenings. Through radio, the shortwave listener becomes familiar with neighborhoods (countries) in the world community.

You could never reach that kind of awareness in a classroom, nor find it in a book, nor attain it any other way, except by extensive travel. That's why you so commonly hear the phrase "tune in the world" connected with shortwave listening.

World-community politics is not all you can hear on shortwave radio. Drama and excitement accompany radio messages between ships at sea and to their shore stations. Maritime companies all over the world conduct business through shortwave radio. The U.S. Navy uses ship-to-ship and ship-to-shore shortwave in its daily activities. Passengers on ocean liners and yachts call their home or business through radio telephone. You can hear them on shortwave.

A planetwide community needs global communications. High-powered long-range radio communication systems carry messages around the Earth. In approximately one-fifteenth of a second, a radio transmission from anywhere on Earth reaches the point farthest away. That's how a businessman in New York or a politician in Washington converses with his counterpart in Tokyo, Peking, or in any other world city.

Atmospheric disturbances occasionally mess up ordinary radio communications. Yet, true communication depends on understanding what is said. Space technology and satellites have moved earthwide communications a step nearer perfection.

The Intelsat series of communications satellites "park" in earth-synchronous orbit. That means they orbit in exact synchronization with the rotation of the Earth. This leaves them stationary above certain locations on the surface. The extreme altitude lets a satellite "see" far beyond the normal radio horizon. Messages go up from one country and beam down to another far away. Special frequencies, impervious to atmospheric disturbance, assure the dependability of communications.

The well-being of any community is tied to its communications. The people need to know what is going on. Whether their movements encompass only the city they live in, or the limits of the earth, there is no room nowadays for ignorance or detachment. People must converse. Citizens must talk things over. They have mutual problems to discuss. Lines of communication must stay open.

On the local scale—that is, in city, state, or country—wired communications dominate. The telephone keeps you in touch with your neighbors. Two-way radio plays a secondary though important role in short-range and regional communication.

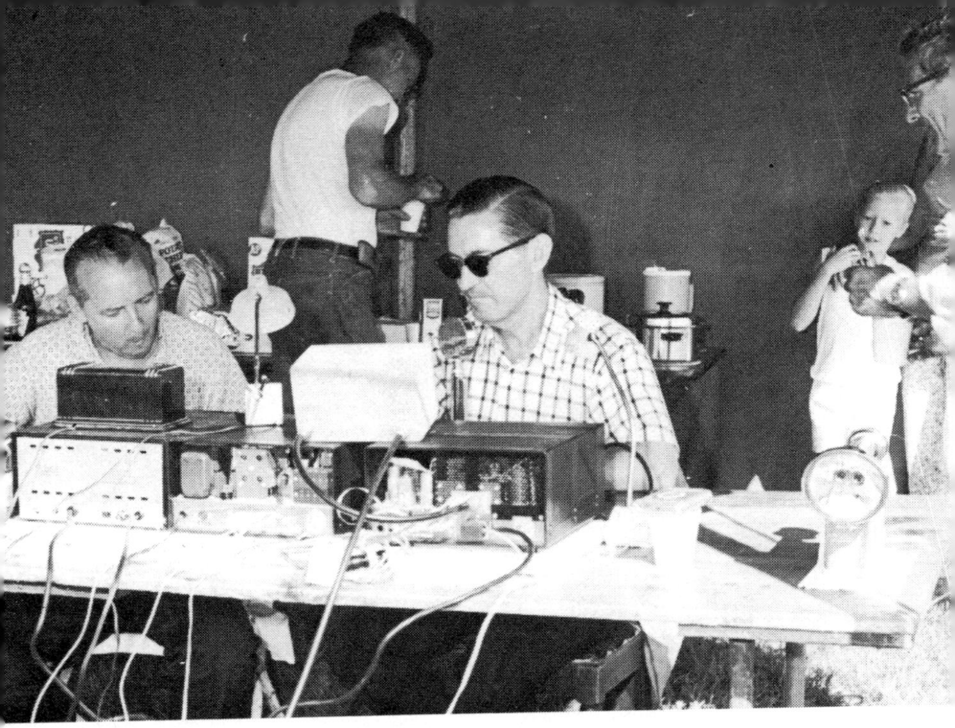

Many shortwave listeners like to pick up Amateur Radio Service broadcasts. These are guys who use radio communications for a hobby. They are often called *hams.* They talk with each other on their own bands of frequencies (listed in Chapter 10.) They not only chit-chat with each other about everything under the sun, but organize field days and jamborees that make great listening. Some hams use a special kind of broadcasting called *single sideband.* To hear them takes a receiver that handles ssb transmissions. A lot of interesting ham messages go by cw (continuous wave) transmission in Morse code; you might find yourself yearning to understand the code.

When isolated areas are stricken by tornado, flood, or other natural disasters, hams generally handle most of the messages into and out of the stricken area. That's fascinating listening. In the photo, hams from one local group, the Delmont Radio Club, test their emergency preparedness with a field trip and actual operation. They communicate with other ham operators to sharpen procedures and to make sure their equipment is ready for a disaster operation.

Our armed forces are scattered all over the world community. We keep in touch with them by radio. Troops in combat, of course, use two-way radios but they are short-range devices. In many ways they resemble what our police and firemen use in our own towns.

For long distances, messages to and from combat troops go in Morse code. They are further coded to prevent interception and understanding by an enemy. (Sometimes, we keep secrets even from our friendly neighbors.) The likelihood of your listening in on these military broadcasts is rather small.

Yet you can listen to certain broadcasts directed to our armed forces. They orginate with the American Forces Radio and Television Service, operated by the Department of Defense.

Powerful transmitters beam regular broadcasts to all parts of the world. The programs consist of news and music to bring "a bit of home" to our men overseas. Of course, since they originate from our section of the world community, these broadcasts carry our point of view. Listening can bring you a deeper understanding of what we tell our troops while they are away.

It's interesting sometimes to compare news broadcasts that go out over AFRTS transmitters with those here at home, on our own radio or television sets. Another fascinating comparison: AFRTS broadcasts to, say, Germany, alongside what you hear beamed back in this direction by the Germans. This is one sure way to learn about world community viewpoints.

Possibly one of the best-known shortwave operations of the United States is Voice of America. Through these facilities, the United States Information Agency presents the way we feel about world affairs to citizens of other sectors. Voice of America sends out programs regularly in 35 major languages, and in other languages for special occasions or purposes.

The master control facility for VOA is in Washington, D.C. There, 23 studios originate the broadcasts and programs that are transmitted throughout the world. More than 1350 people work there and at four transmitting facilities scattered around the United States.

Actually, there are 41 VOA transmitters in this country. One location is near Bethany, Ohio; two are in California, at Delano and Dixon; one at Marathon, Florida; and one at Greenville, North Carolina. Antenna systems at these facilities beam shortwave broadcasts overseas. There, special relay transmitters in turn rebroadcast the programs in a direction to blanket the countries for which they are intended. Photo shows the VOA relay station at Tinang in the Philippines.

There are 68 transmitters at VOA facilities in other countries. We have installations in Ceylon, Thailand, South Vietnam, the Philippines, Morrocco, Liberia, England, West Germany, Okinawa, and Greece. Programs from these transmitters are beamed into specific large areas: Southeast Asia, the Middle East, Southern China, the Soviet Union, Africa, Latin America, and the Caribbean countries and islands. Broadcasts utilize the principal language spoken in each target segment of our world community. In some instances, announcers repeat material in local dialects to make sure everyone in a locality has a chance to hear and understand.

The Voice of America broadcasts nearly 800 hours per week. You can get seasonal schedules by writing to Broadcasting Service, U.S. Information Agency, Washington, DC 20547. The schedules tell where each broadcast is aimed, time of day for the broadcast, the frequency (or frequencies), and which facility in the United States originates the broadcast.

Remember that only some programs are transmitted in English. If you happen to speak or study some foreign language and wonder what we're saying to natives of that country, tune in the VOA broadcasts aimed there.

Some broadcasts are regular. Others are special broadcasts, of a documentary nature. News broadcasts are the most common programs. For example, in the photo on this page, broadcasters prepare to report one of our Apollo moon flights. This particular broadcast went to West Pakistan; the language was Urdu.

Programs from Voice of America transmitters reach at least 50 million adults over age 14. In addition, some 4,000 local radio stations in various countries broadcast taped programs from VOA or rebroadcast certain shortwave transmissions.

One of the most popular programs is called "Music USA." You might be interested in tuning in this one yourself. The time of day varies according to destination of the program. Consult the latest schedules to be sure of time and frequency.

Also, broadcasts in very simple English are made to certain countries. The program vocabulary consists of only 2000 most-used words. Thus, the peoples of the world can hear news, scientific information, and certain types of cultural broadcasts, in English.

Scenes like this are typical for Voice of America reception. However, it works in reverse too. You could see the same thing in many living rooms in the United States. Immigrants from another part of the world community like to keep in touch. Having been accustomed to a different culture, they turn to shortwave radio to bring music and news from "home."

Americans interested in a particular country, or who have relatives living there, sometimes listen to overseas broadcasts. Or, some might simply like the music of another country. Any time you have an urge to zero-in on something unusual, try listening to music from some of the Asian cultures. Likewise, certain South American stations have music programs that are different from anything you can hear locally.

Not all shortwave listening relates to foreign countries. For example, right in your home city, one popular pastime is listening to public-safety channels.

A two-way radio-communications center has responsibility for the safety and well-being of citizens for many miles around. With an appropriate receiver, you can hear messages to and from police, ambulance, and fire vehicles. You may have to fish around with the tuning dial to find these signals. Or perhaps you know someone who can tell you the local frequencies.

You'll learn what goes on around your city or county. Public-safety listening gives you a new viewpoint about crime in your area. It can also impart an awareness of the many services rendered by your police and fire departments.

The special shortwave receiver (page 70) that picks up these services can also introduce you to taxicabs, concrete trucks, and a myriad other business communications that go on all day. Just spend a little time hunting up and down the dials for the two-way communications that interest you.

Shortwave signals that carry police and fire communications differ considerably from the kind that go around the world. You can't pick up the former on an ordinary shortwave radio. These local signals are called very high frequencies—abbreviated *vhf.* In fact, some police and fire departments have shifted to ultra high frequencies or *uhf.*

Furthermore, vhf and uhf two-way radios use a mode of transmission unlike overseas shortwave. Whereas ordinary shortwave broadcasts use a-m (amplitude modulation), public-safety services use fm (frequency modulation).

Some police and fire departments frown severely on citizens listening to their communications. However, you have a right to do so. But you are forbidden under federal law to repeat anything you hear or to use it for your own benefit. Hence, if you listen to police calls to avoid being caught in a crime, you violate both federal and local laws.

Here is something you might hear from on more than one kind of receiver. Since the helicopter is a form of aircraft, it carries radio transmitting/receiving equipment similar to that installed in any other plane.

But this is a police helicopter. It also totes police two-way radio so it can communicate with headquarters and with police vehicles on the ground. For traffic control and search operations, the police helicopter proves to be one of the handiest tools. It consequently provides some of the most interesting police-band listening.

The airliners of the world depend on radio to keep them in touch with traffic controllers, with their company at home, and with each other as they fly the air lanes. They use many different frequencies, depending on which part of the world they are heading to or from. Sometimes the time of day or night has a bearing on exactly what frequency they use.

But you can listen. With one kind of shortwave receiver, you can hear planes far out over the Atlantic or Pacific oceans. With another, you can hear them preparing to land at the airport near your city. Their altitude lets you hear some conversations from a hundred miles away. For example, if you live in New Jersey, you can hear the conversation of pilots of planes landing at Kennedy, LaGuardia, Newark, Philadelphia, and sometimes Baltimore.

Occasionally, you can hear long-distance phone calls placed from airliners.

Around a busy airport, planes must communicate with a Federal Aviation Administration (FAA) control tower. This makes fun listening, if you have an affinity for aircraft or flying.

Any Sunday afternoon, near an airport with many planes flying, you can hear all sorts of communications between aircraft and ground. Or, you can learn the tower frequency of your metropolitan airport and enjoy hearing large airliners arrive and depart.

Aircraft shortwave listening demands still another kind of receiver. Aircraft communications take place in the vhf and uhf regions—principally vhf. So, to pick them up, a receiver must tune these specific frequencies.

But unlike police vhf and uhf transmissions, aircraft use the a-m (amplitude modulation) mode. Chapter 4 acquaints you with many different types of receivers, including this kind.

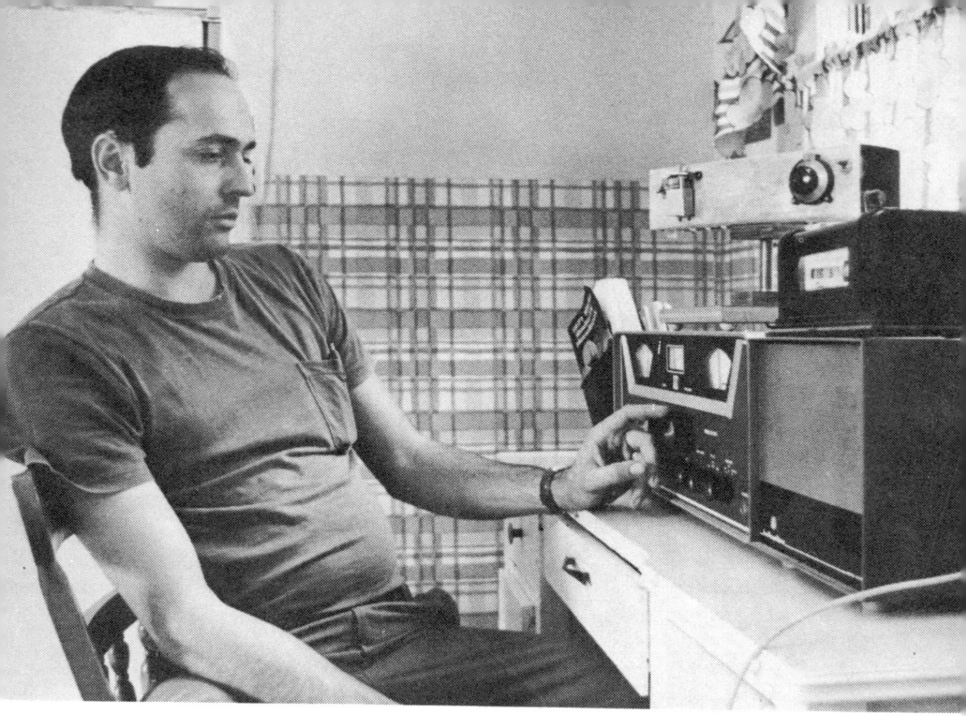

So that's what shortwave listening amounts to. Whether your interest lies in what goes on in your own locality or in the global community, shortwave receivers open the doors for you.

Who listens to shortwave? Anybody can. All you need is an interest, a receiver, and a little time to spend twirling the dial; mix in sincere curiosity about what's happening in the world around you. In the pages that follow, you'll see ways other people have fun listening to shortwave radio.

Chapter 2

The Language of Shortwave

Every group talks its own special language. So do shortwave listeners. But this tendency has nothing to do with snob appeal. Radio communications generates a certain amount of technical terminology. If you want to enjoy shortwave radio, you should learn the vocabulary. Even if you don't join a group, knowing your equipment well necessitates familiarity with certain technical terms. The following pages introduce you to words and concepts that relate to shortwave listening.

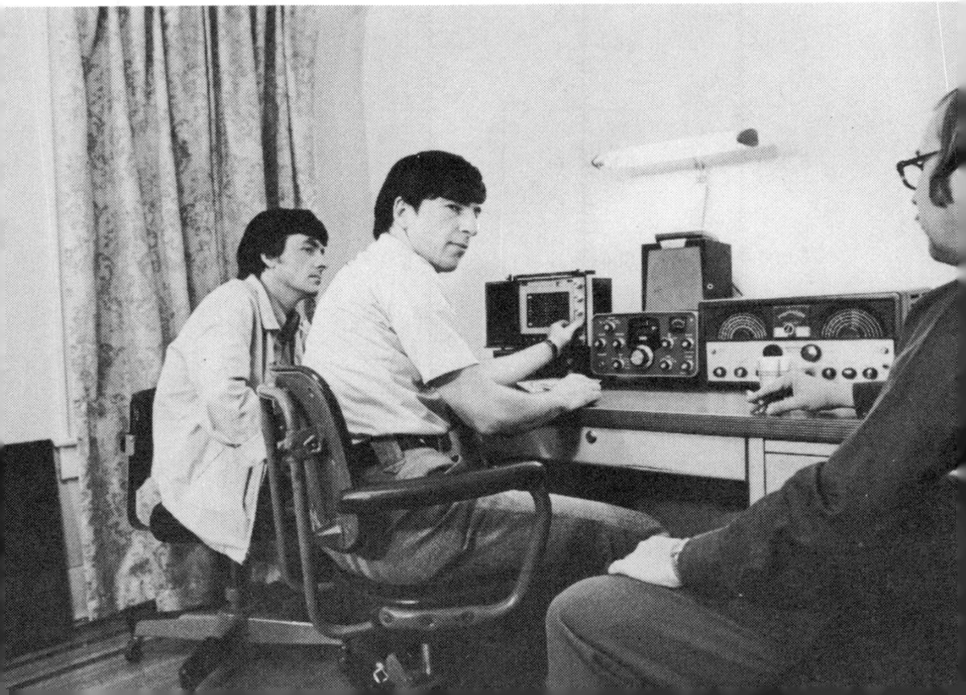

Frequency-Wavelength Conversion Chart

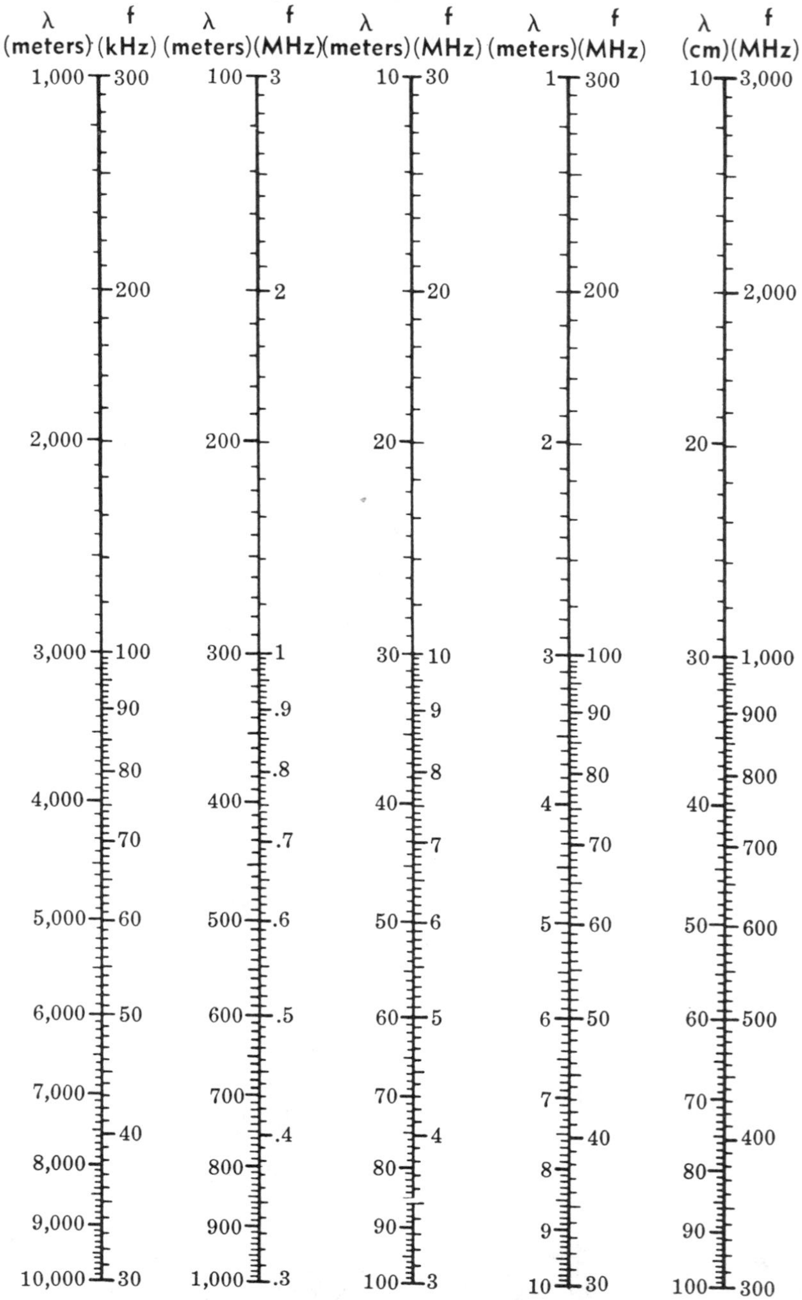

Two words you'll encounter first may seem interchangeable. They're not. In fact, one is exactly the reciprocal of the other, mathematically speaking. The two words are *frequency* and *wavelength.*

Consider wavelength first. Radio waves, as they travel through the air, have certain dimensions. Some are comparatively long and are called *long waves.* Others, not so long, are *medium waves.* Those waves whose length is even less are called *shortwaves.*

By international agreement, radio wavelengths are measured in meters rather than in yards or feet. Transmitters that put out radio waves one thousand meters long are said to operate at a wavelength of one thousand meters. A 10-meter station transmits radio signals in which each wave is only 10 meters long.

Radio waves are *transverse waves,* with peaks and dips. The length of a wave is the distance between any two consecutive peaks (or any two consecutive dips). As radio waves travel at a speed of 300,000,000 meters per second (about 186,000 miles per second), the peaks and dips of a radio signal pass any given point at that high speed. With the proper equipment, you can measure how many peaks pass that point per second. The number of peaks measured is called the *frequency* of the signal. Frequency is expressed in *hertz.* (Older books and charts used the term *cycles per second.* One peak and one dip is one cycle. One cycle per second is one hertz—abbreviated Hz.)

If you are mathematically inclined, you can see a relationship between radio frequency, velocity, and wavelength. Dividing frequency into velocity gives you wavelength; dividing wavelength into velocity gives you frequency. For example, consider 30 million peaks (and dips) passing a specified point each second; the radio wave is traveling at a speed of 300 million meters per second. Thus, the distance from one peak to the next peak (length of one radio wave) is 10 meters. Therefore, a 10-meter radio wave has a frequency of 30 million hertz or 30 megahertz —abbreviated MHz.

As another example, the wave peaks of the 600-kilohertz (kHz) broadcast station develop at a rate of 600,000 per second. (Kilo means one thousand.) Frequency is 600 kHz and the velocity is constant at 300 million meters per second. The wavelength (distance between peaks of its waves) is 500 meters. (Velocity of 300 million divided by frequency of 600,000 equals wavelength of 500.)

Earlier pages mention long wave and shortwave. There are two reasons to know what these words mean. First, frequencies are grouped as *long wave, medium wave, shortwave,* etc. Second, the wavelength of the radio signal you want to receive determines what kind or how long an antenna you should use. Any radio wave approximately 600 meters or more between peaks is considered a long wave, thus any frequency below about 500 kHz is a long-wave signal. (The definite division between the length of a long-wave signal and the length of a medium-wave signal depends on the reference book you look at.)

A few aircraft and ship navigation beacons and weather stations still operate at frequencies below 400 kHz. Certain naval communications take place even lower than that. Generally, however, no one listens to long-wave stations very much. You can get a receiver that has the long-wave band in case you want to hear weather broadcasts in your vicinity.

You can often tell by its antenna whether a radio station is used for medium-wave or long-wave transmission. Long-wave stations require lengthy antennas for transmitting.

Medium-wave stations, the kind that usually serve your town and are called a-m stations, can have a *vertical radiator.* That is, they can put up a tall tower and let the tower itself be the transmitting antenna.

Medium-wave stations are considered those that operate from about 500 kHz to 2000 kHz (2 megahertz). Among some radio "experts," only a-m broadcast transmissions are classed as medium wave. That situates medium-wave stations between the frequencies of 540 and 1600 kHz. A few medium-wave stations operate at frequencies near 2000 kHz in Europe.

Then there's the word for which this book is named: *shortwave*. These waves are those above 2 MHz, with a wavelength of 150 meters or less. Shortwave frequencies go to about 30 MHz. Here, the waves are only 10 meters long. Again, there is some slight discrepancy as to exactly what frequencies are called shortwave. Officially, shortwaves are those that measure from as long as 100 meters (at 3 MHz) to as short as 10 meters (at 30 MHz).

Even the shortwave sector of the radio spectrum is divided sometimes. Frequencies at the lower end take long-wire antennas. That is, a wire is strung between two insulators on trees or poles. A lead-in wire is attached between the antenna and receiver to carry incoming radio signals to the shortwave receiver. When frequencies approach 30 MHz (for example, CB radio frequencies are near 27 MHz), antennas can be short enough to be mounted on a tower and still be efficient. The crossover point between the two types of antennas "divides" the shortwave band.

Beyond 30 MHz is a frequency range called *very high frequency.* It's abbreviated *vhf.* At these frequencies, the wavelengths are so short they seldom are mentioned, except in two cases. Frequencies around 50 MHz have wavelengths about 6 meters long. Frequencies at about 150 MHz have wavelengths approximately 2 meters long. Certain ham radio stations are authorized to use frequencies in these two ranges. It's easier to pronounce "2-meter band" or "6-meter band" than it is to say "50-million-hertz band" or "150-megahertz band."

Officially, vhf begins at 30 MHz and extends to 300 MHz. Beyond that, from 300 to 3000 MHz, the frequency range is called *ultrahigh frequency* or *uhf.* Both vhf and uhf take special receivers. The ordinary shortwave receiver won't pick them up.

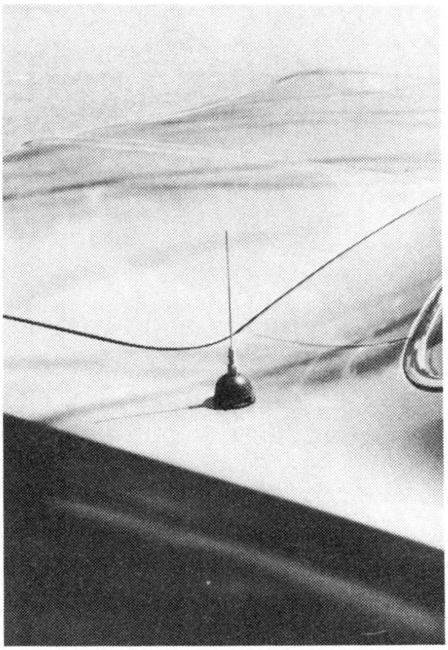

Furthermore, they take short antennas. The longer one here is cut to one-fourth wavelength; it's a quarter-wave antenna for a frequency of 150 MHz. The short one is cut to a quarter wavelength for 450 MHz.

Many terms in shortwave listening go by their initials. Typical of these are SSB, USB, LSB, and DSB. The letters common to all these are S and B. They stand for *sideband.*

The process of putting a voice on a radio signal is called *modulation.* Ordinary (amplitude) modulation creates *sideband frequencies* adjacent to the main carrier frequency. There's one sideband just above the carrier frequency, another just below. Hence the name *double-sideband* or *DSB.* A receiver picking up the signal demodulates the sidebands to recover the voice.

For efficiency, one sideband and the carrier can be dispensed with. That's what happens when a transmitter operates in the SSB or single-sideband mode. USB stands for *upper sideband* (lower sideband and carrier eliminated). LSB means *lower sideband* (upper sideband and carrier eliminated).

Another important letter designation is CW. This means *continuous wave.* What CW actually refers to is an *interrupted continuous wave* (ICW).

A radio signal transmitted as a series of long and short pulses is called a code transmission. The pulses are sent in a special pattern known as Morse code. Many shortwave listeners, and shortwave manufacturers, erroneously call code reception "CW reception." The receiver mode switch for code reception may be marked with a simple CW, or with CWO or BFO.

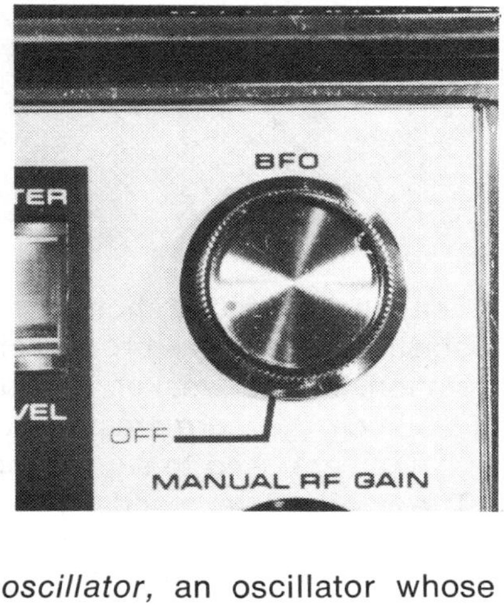

The latter two designations refer to the means by which a receiver picks up code. A little oscillator inside the receiver turns on and "beats" with the incoming interrupted-continuous-wave signal. The oscillator is close to an intermediate frequency in the receiver. The proximity of the two signals causes an audible note to be heard in the speaker or headphone of the receiver. The sound comes in as dits and dahs that conform to the transmitted ICW.

BFO means *beat-frequency oscillator,* an oscillator whose signal beats against the incoming signal to create an audible note or sound (at the speaker).

The goal of most shortwave listeners, at least those who specialize in overseas broadcasts, is distance. The further away a station, or the weaker its transmitting power in relation to its distance, the more prestige accompanies receiving it. The letters "DX" are used to signify reception over a long distance.

The D, of course, stands for distance. The X comes from communications shorthand for transmit (xmit), transmitter (xmtr), and transmission (xmn). Thus DX is long-distance transmission. If you're at the listening end, DX means you are receiving a broadcast from far away.

Even though you keep a careful listing (log) of each station you receive, that's not considered proof. Not that anyone would accuse you of dishonesty, but a station might fool you. The calibration of your receiver dial might be slightly inaccurate, you might misread it, or the dial may be too broadly marked for precise determination of a station's frequency. You might misunderstand the identifying call sign of a foreign station. The real DX prizes are usually difficult to sort out from among other stations operating on the same or closely adjacent frequencies.

But you can get proof that you heard a certain station. Write to the station, saying that you received them. Mention the date and approximate frequency, and describe what you heard and the listening conditions. Ask the station to verify that they were indeed transmitting that program material at that moment.

The typical verification comes in card form, called a *QSL card.* QSL comes from Morse-code shorthand for "I am acknowledging that I received your message (or that you received mine)." Some verifications are letters. In whatever form, serious shortwave listeners take much pride in their QSL collections.

Shortwave broadcasting stations typically give time-of-day in *Greenwich* (pronounced Grin′ ich) *Mean Time* (GMT). It's the time of day at the earth's prime or zero longitudinal meridian, which happens to bisect Greenwich, England. GMT is the time standard for the entire world community. Since Earth takes 24 hours to rotate once, the time at Greenwich goes from 00:00 to 24:00.

Other terms are used for GMT occasionally. It's called World Time, Worldwide Time, Universal Time, Meridian Time, 24-Hour Time, and—by pilots—Z-Time or Zebra (or Zulu) Time.

The use of GMT for shortwave schedules facilitates communications. You see, 21:30 GMT occurs at exactly the same time in the United States as in the Soviet Union, or in Antarctica, or anywhere else in the world. This time designation even works for outer space, assuming Earth is the point of reference.

To convert GMT to local time anywhere in the world, you simply add or subtract the time difference between "sun time" and GMT.

The eastern United States gets the sun about four hours after Greenwich, England. Therefore Eastern Standard Time (EST) always lags GMT by that four hours. When GMT is 07:00, EST is 03:00. (Time in the 24-hour system is usually written without the colon, which is how it will appear hereafter in this book.) At 1200 EST, which is noon in New York, world time is 1600 GMT.

When a shortwave broadcast in Moscow goes on the air at 2145 GMT, the Muscovite announcer's watch says 45 minutes past midnight (0045) because Moscow sunset comes three hours before sunset at Greenwich. The kitchen clock of a listener somewhere in Colorado points to 3:45 in the afternoon—or 1545 MST (Mountain Standard Time) on the 24-hour system.

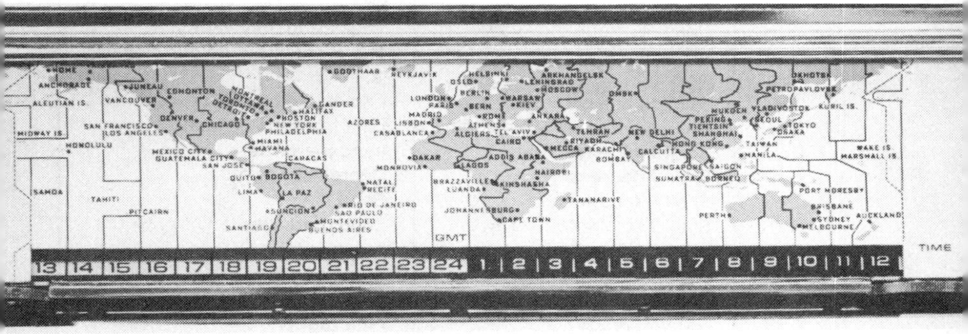

In the illustration, you can count the number of hours from the prime meridian to any part of the world. *Add* that number of hours to GMT if that part of the world is east of Greenwich, England; *subtract* if the location is west of the GMT time zone. (If you're on Daylight Saving Time, subtract an extra hour from GMT.)

In conversations among hams, you may hear phrases like "There's too much *QRM*" or "I'll *QRT* till tomorrow night same time." You may find similar three-letter groups in code messages you copy.

These letters are shorthand for certain often-used phrases or messages. Each has a declarative and interrogative meaning, the latter being followed by a question mark. Some popular *Q signals* are shown below.

Hence, *QSO?* in a code message would mean the operator at the transmitting station is asking if the receiving operator knows how to reach some third party. If he knows a way, the return message might be QSO STATION X VIA RLY THRU WZ4ZZZ. That's to say station X can be contacted by asking station WZ4ZZZ to relay the message. The *Q signals* save time and speed up code communications.

Signal	Question	Answer or Advice
QRG	Will you tell me my exact frequency?	Your exact frequency is . . . . kHz (or MHz).
QRH	Does my frequency vary?	Your frequency varies.
QRK	What is the readability of my signals?	The readability of your signals is . . . .
QRM	Are you being interfered with?	I am being interfered with.
QRN	Are you troubled by static?	I am troubled by static.
QRO	Shall I increase power?	Increase power.
QRP	Shall I decrease power?	Decrease power.
QRQ	Shall I send faster?	Send faster.
QRS	Shall I send more slowly?	Send more slowly (. . . . words per minute).
QRT	Shall I stop sending?	Stop sending.
QRU	Have you anything for me?	I have nothing for you.
QRV	Are you ready?	I am ready.
QRX	When will you call again?	I will call you again at . . . . hours [on . . . . kHz (or MHz)].
QSA	What is the strength of my signals?	The strength of your signals is . . . .
QSB	Are my signals fading?	Your signals are fading.
QSL	Can you acknowledge receipt?	I am acknowledging receipt.
QSM	Shall I repeat the last message I sent you?	Repeat the last message you have sent me.
QSO	Can you communicate with . . . . direct or by relay?	I can communicate with . . . . direct (or by relay through . . . .).
QSV	Shall I send a series of V's?	Send a series of V's.
QSY	Shall I change to transmission on another frequency?	Change to transmission on another frequency [or on . . . kHz (or MHz)].
QSZ	Shall I send each word or group twice?	Send each word or group twice.
QTH	What is your location?	My location is . . . . .

Chapter 3

Frequency "Bands" and Their Users

Until you learn to sort out the various numbers, the dial of a multiband shortwave receiver is a maze of confusion. But the dials of all modern receivers, whether for long-wave, medium-wave or shortwave reception, are marked with frequency numbers.

Always, the number represents frequencies. Shortwave stations are lumped together in groups of frequencies or *bands.* Some of the groups, as you'll see, are named for their wavelength: 31-meter band, 16-meter band, 2-meter band, etc. The ensuing pages familiarize you with some of the most popular bands and communication services.

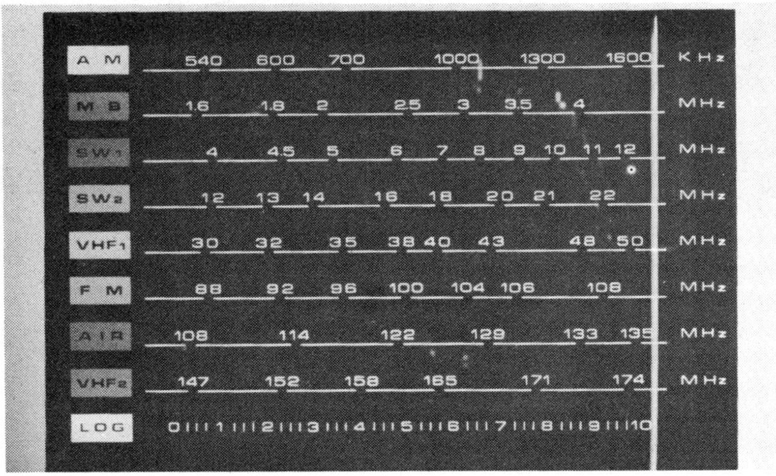

Not many receivers can tune long-wave frequencies and there isn't much to hear on long waves anymore. Only a few sea/air navigation beacons and weather stations now use this band. These services are scattered between 190 and 400 kHz in frequency. The receiver shown here tunes down to 150 kHz, where a few naval transmitters operate. Most transoceanic navigation services have been moved into high frequencies (shortwaves) or to satellites.

The long-wave international distress frequency at 500 kHz can be picked up only by special alarm receivers. Ships at sea and transoceanic aircraft flights monitor this frequency, but only transmit on it in case of dire emergency. The whole long-wave spectrum has fallen into disuse in favor of more dependable higher frequencies.

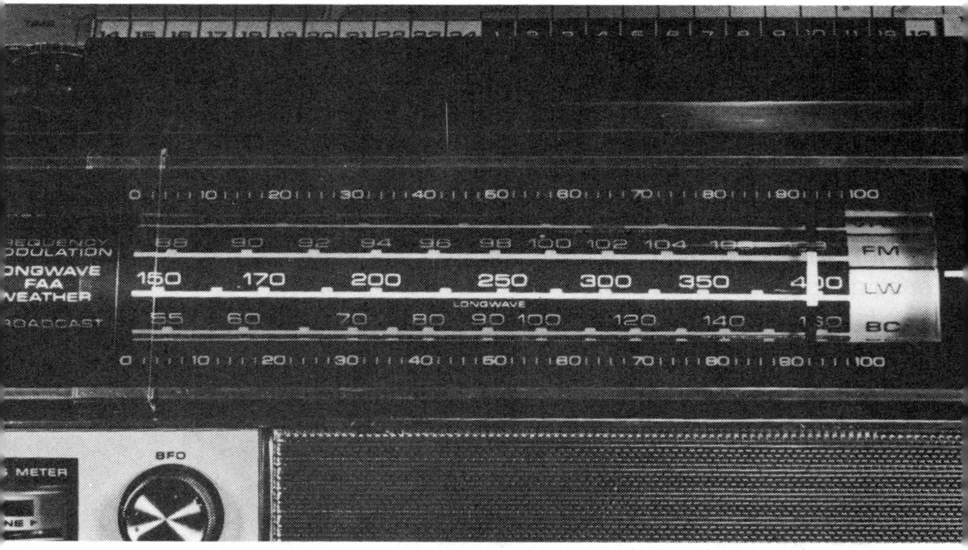

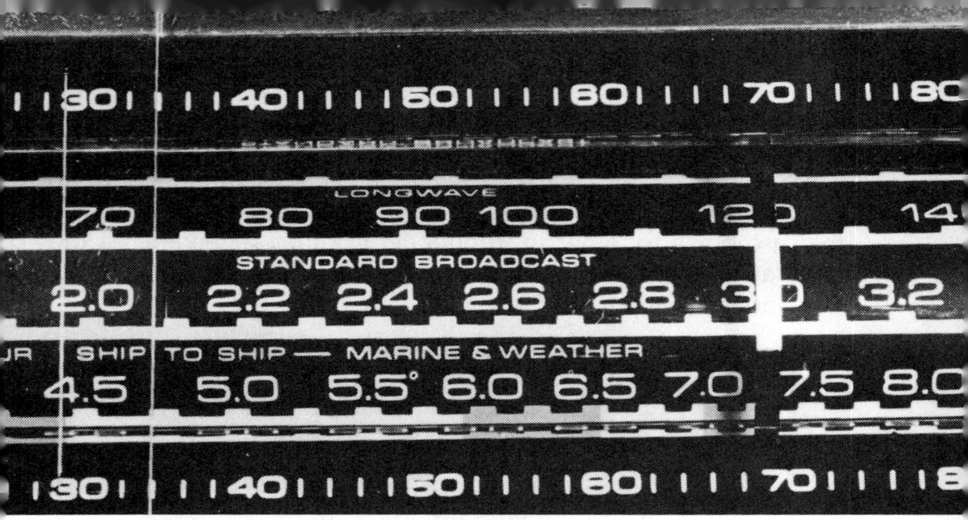

The band of frequencies from 2 to 3 MHz (2000-3000 kHz) used to be popular among shortwave listeners. At night, stations using these frequencies can be heard for thousands of miles. That's a problem; the ordinary shortwave receiver has trouble spreading out the high-density crowding. In daytime, the frequencies are useless except for distances under 100 miles.

Ships at sea, particularly along the coasts, and some aircraft use these frequencies. On the Great Lakes, weather bulletins are transmitted regularly on frequencies in this band.

Technically, 2-3 MHz is medium wave. But habit among communications people has tagged this as a shortwave band. Usually, it's called "the 2- to 3-MHz marine band" because river and lake boats use several frequencies within the band.

One distress frequency lies within this band. Sometimes at night, you hear traffic at 2182 kHz. Usually, you can hear so many ships, you can understand no one. Very soon, virtually all these transmissions will be ssb.

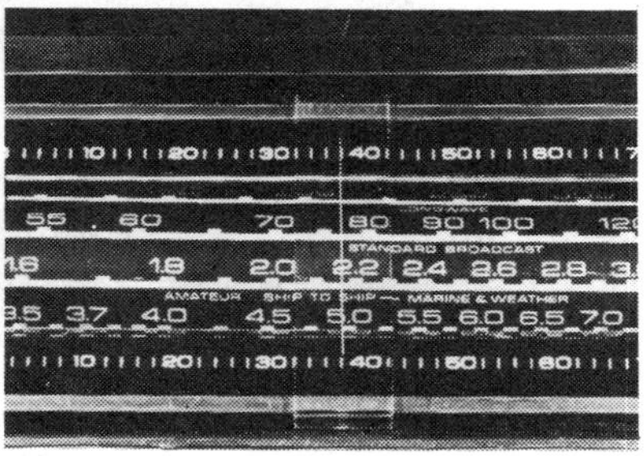

That brings you to what are called the "international shortwave bands." A few radios spread these small segments of radio spectrum out over a wide dial. That helps you separate stations that are close together in frequency. In other receivers, as the illustrations on this page demonstrate, a "band" may be only one narrow segment of the dial.

The 49-meter international shortwave broadcast band encompasses frequencies from 5.95 to 6.2 MHz. These frequency allocations are determined by treaties, and are adhered to by virtually all countries.

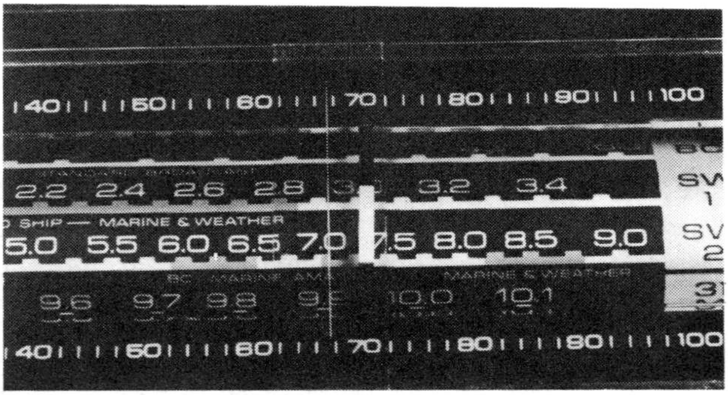

The 41-meter band lies from 7.1 to 7.3 MHz. As you'll see later, this falls right in the 40-meter ham band. What you hear on these frequencies, on a particular evening, depends on how powerful the transmitters are. Shortwave broadcast stations usually win out, because hams are limited to 1000 watts. The interference of stations with each other (QRM) is sometimes so heavy that you'll dial elsewhere anyway.

The 49- and 41-meter international broadcast bands are not the most popular, partly because they are heard well in the United States only at night. Their frequencies don't carry far in the daytime. Stations you hear before dark on these bands are either nearby (Latin America) or extremely powerful.

Higher-frequency stations can be heard best during early evening. Some stations are received throughout the day. The 31-, 25-, 19-, 16-, and 13-meter bands are the popular international shortwave broadcast bands.

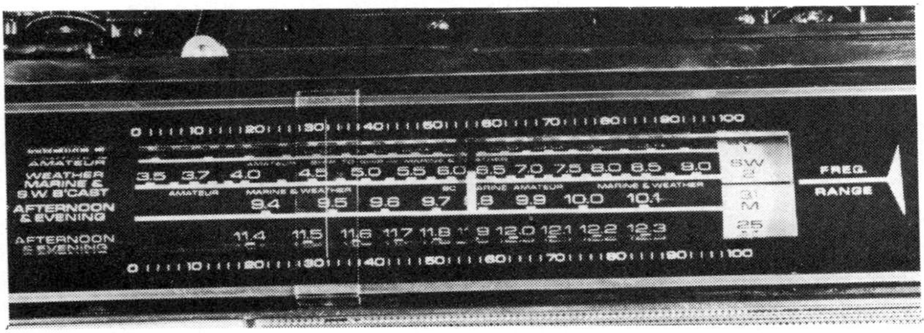

The 31-meter band covers frequencies from 9.5 to 9.775 MHz. That's a pretty narrow spectrum. As you can see, spreading one whole megahertz over the dial still confines the 31-meter band to only a couple inches of numbers. The marker indicates the bottom end of the band; the dial pointer rests at the top.

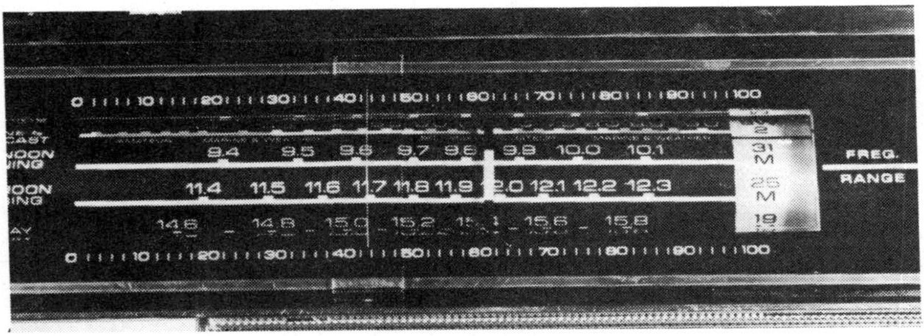

The 25-meter band, being a bit shorter in wavelength, is likewise higher in frequency. Frequencies lie between 11.7 and 11.975 MHz. Again, a narrow-spread dial marking is used, covering one megahertz from 11.3 to 12.3 MHz. This makes tuning easier than it would be on a radio with an ordinary multifrequency dial (covering several megahertz).

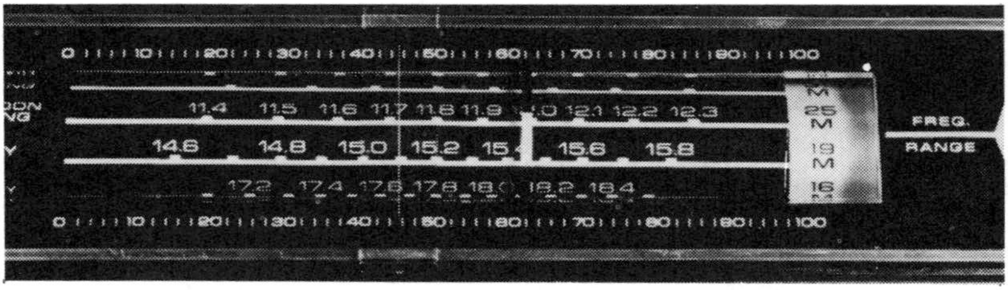

The 19-meter shortwave broadcast band covers frequencies from 15.1 to 15.45 MHz. Notice that the higher the frequencies go, the harder it is to spread the band out over a wide segment of dial. You have to tune more carefully.

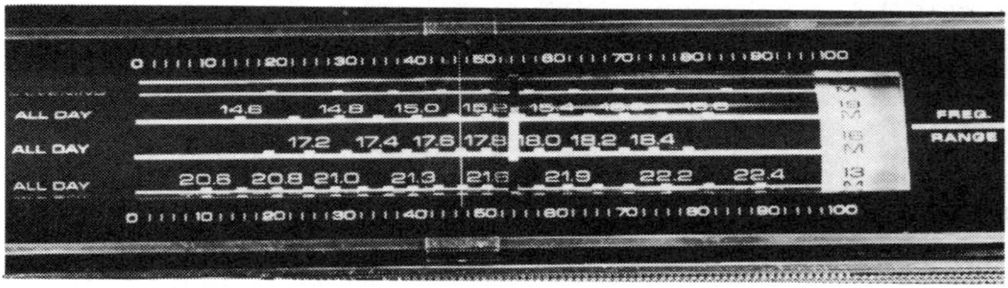

The 16-meter band reaches from 17.7 to 17.9 MHz. The spread on the dial is very narrow. This band is popular for shortwave stations wanting to reach into the United States around noontime; the frequencies propagate well during the daytime.

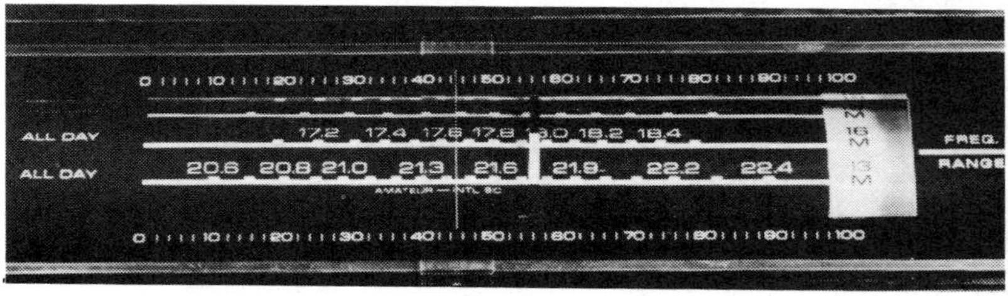

Frequencies in the 13-meter band travel excellently throughout the world in the daylight hours. All the transmitting stations need is a reasonably strong power. You can tune these broadcasts between 21.45 and 21.75 MHz. The band generally isn't crowded, but turn the dial knob slowly and patiently.

The least popular international shortwave bands lie in the 2- to 5-MHz range. They are useful between neighboring countries and are put to that use. The stations don't ordinarily supply the transmitter power necessary to push these comparatively longer wavelengths around the world. Frequencies used are:

120-meter band—2.3 to 2.495 MHz
90-meter band—3.2 to 3.4 MHz
75-meter band—3.9 to 4.0 MHz
(shared with hams)
60-meter band—4.75 to 5.06 MHz

You can try them if your receiver tunes these frequencies. The most likely time is at night, in the United States.

120 M

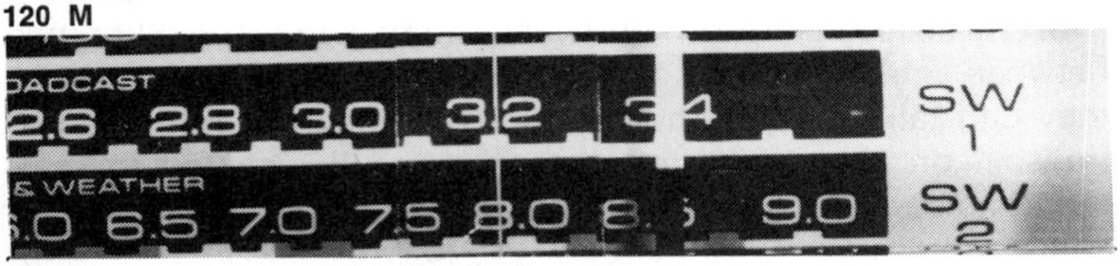

90 M

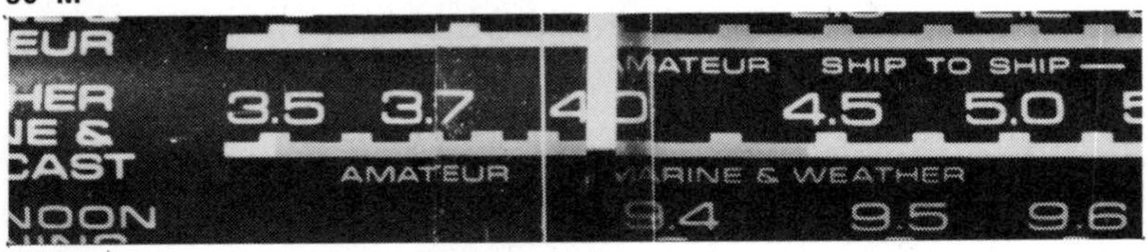

75 M

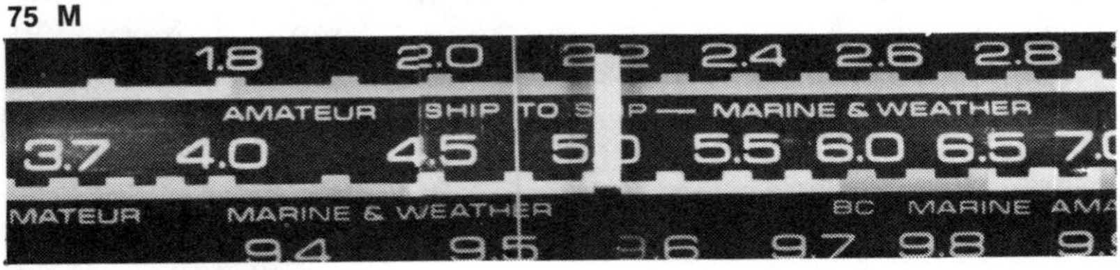

60 M

Amateur Radio Service is the official name for the ham bands. Treaties and our own laws allocate several groups of frequencies to amateur radio communications. They begin at wavelengths as long as 160 meters; they include some frequencies far into the experimental microwave region. For practical listening, the shortest ham wavelength you'll probably be interested in is 2 meters.

In one sense, hams are shortwave listeners too. The difference between you and them is that they can talk back. But of course they can talk only to other hams. Also, they are forbidden to transmit on frequencies other than those designated for amateur radio.

Certain shortwave receivers tune only the ham bands, but bring in those few frequencies with high precision. Here are some of the frequencies involved.

160-meter band—1.8 to 2.0 MHz
80-meter band—3.5 to 4.0 MHz
40-meter band—7.0 to 7.3 MHz
20-meter band—14.0 to 14.35 MHz
15-meter band—21.0 to 21.45 MHz
10-meter band—28.0 to 29.7 MHz
6-meter band—50.0 to 54.0 MHz
2-meter band—144.0 to 148.0 MHz

Hams have access to frequencies much higher, but those require special transmitters and receivers. You'll not likely be interested in ham communications in the vhf and uhf ranges. They are very short range and very sparse, limited mostly to experimentation (which is one important purpose of the Amateur Radio Service).

Citizens Radio Service is the Federal Communications Commission (FCC) name for another band of frequencies that have become extremely popular. Colloquially, this communications service goes by the label "Citizens Band" or just "CB."

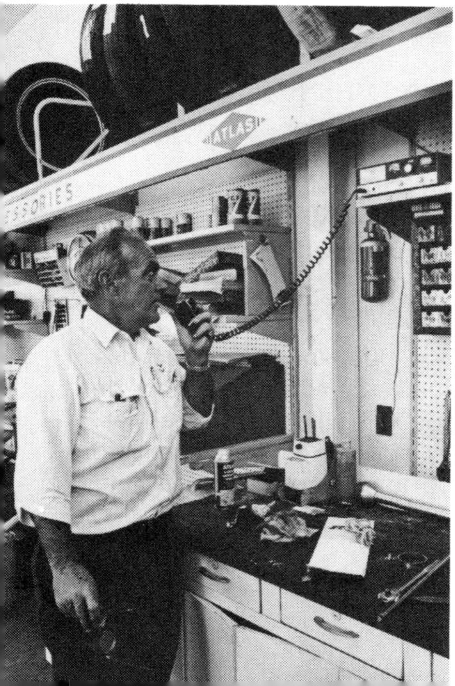

Any U.S. citizen 18 or older can, by filing application and paying the license fee, obtain his own license for two-way CB communications. (*Easi-Guide to CITIZENS BAND RADIO,* published by Howard W. Sams & Co., Inc., explains in detail how to get a CB license, buy the best equipment, and operate a CB station legally.) Citizens Band radio can be used for business or for personal communications, as from home to car, car to car, even from home to home under some circumstances.

Most people interested in Citizens Band radio want to talk both ways. But you can listen to those frequencies too, if you have a receiver that tunes from 26.965 to 27.255 MHz. There are 23 channels of class-D Citizens Band radio. Most good receivers that can pick up the 10-meter ham band can also be tuned for CB listening. You may even find that segment of frequencies already marked on the bandspread dial. (CB stations use a-m or ssb transmission, just like other shortwave stations.)

In all honesty, an awful lot of trivia and foolishness goes on in the CB channels. Many of these operators use their rigs illegally in one way or another. The FCC regularly cracks down on some of the worst offenders, levying fines and revoking licenses. You might enjoy hearing CB transmissions, but it isn't likely you'll pick up any information useful in furthering your hobby of shortwave listening. Listen to the band some evening and decide for yourself.

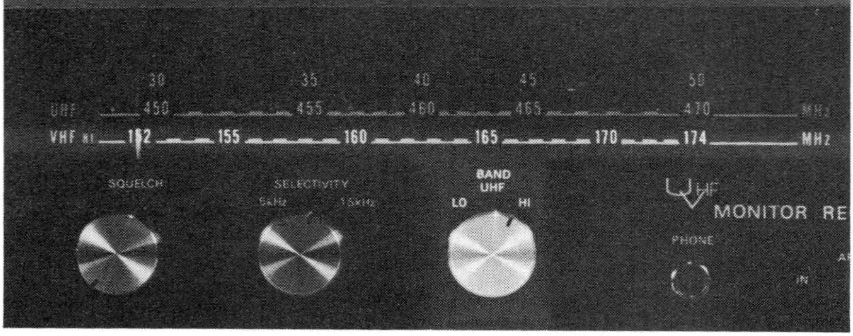

The "two-way radio" bands grow in popularity each year. More offically, these are known as the land-mobile portions of the radio spectrum. Until 1972, only land vehicles used them. Now, a small sector of the "high" vhf band carries short-distance communications (10-30 miles) between ships or riverboats and their shore stations.

There are three major two-way radio bands:

Low-band vhf—30 to 50 MHz
High-band vhf—150 to 175 MHz
Uhf—450 to 470 MHz

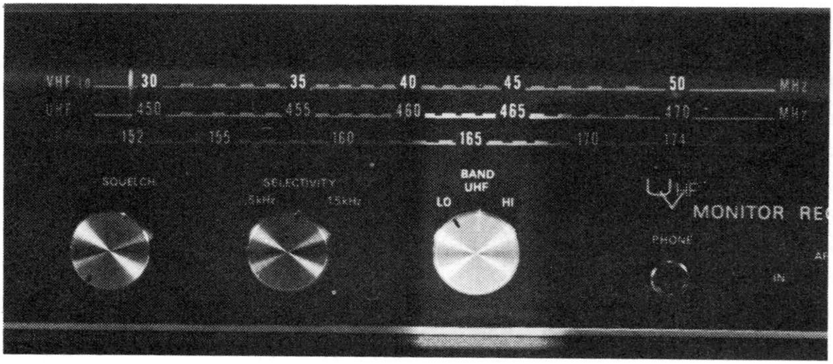

You generally need a special receiver called a *monitor* to hear these communications. Transmissions on all three bands use frequency modulation (fm). This mode is practical and effective at these frequencies whereas it wouldn't be at lower frequencies. (Fm equipment is more expensive, which is one reason CB radio doesn't use that kind of modulation.)

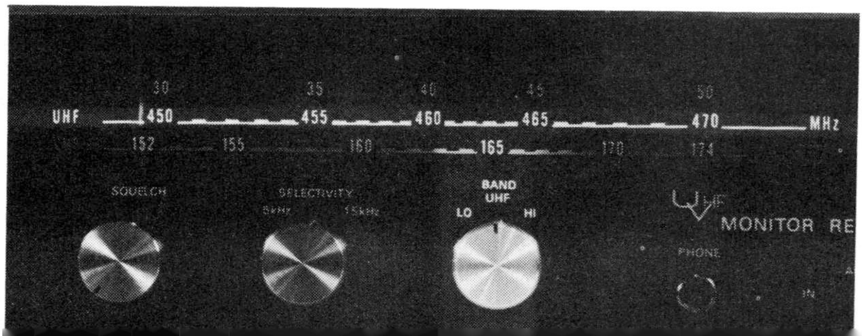

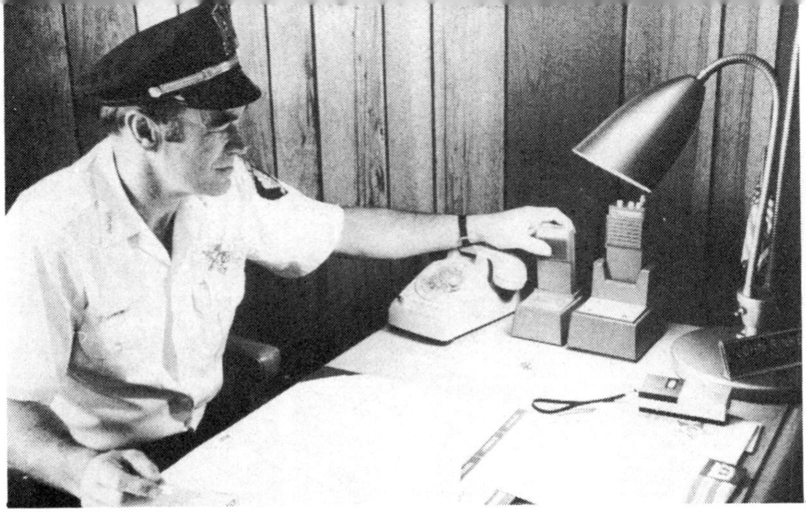

Until lately, the high vhf band has been the most popular. Low vhf seems best for distances beyond 20 miles (up to 40 miles or so, with powerful transmitters and high antennas). State police and sheriff departments have favored that band. So have trucking companies. City police, fire departments, and businesses that need short-range radio communications preferred 150 MHz. That band suffers less from atmospheric disturbances.

Advances in equipment technology have made the uhf band attractive. It's quieter, is less crowded, and can reach just about as far as high-band vhf. Police and fire departments are the first to transfer their communications into the uhf band. So, if you're a serious listener to police and fire signals, be sure any *monitor* receiver you buy can pick up all three two-way bands; older models receive only the two vhf bands.

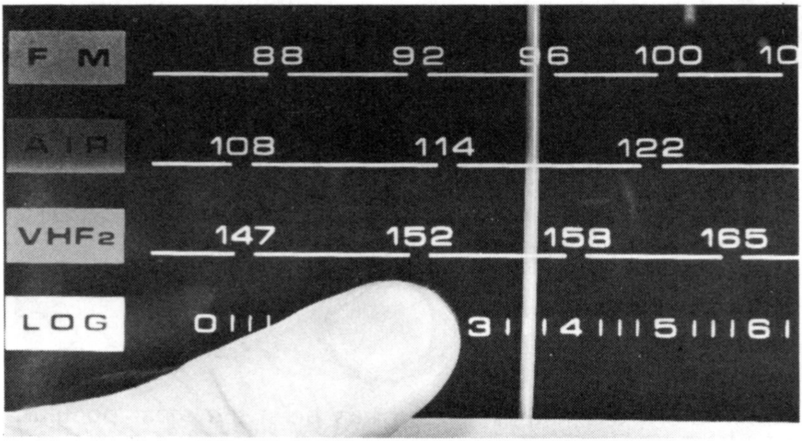

One small segment of the high vhf band, around 156 MHz, carries boat communications. Certain ship-to-shore phone calls are on frequencies up around 162 MHz; but these calls are duplexed and you can hear only one side of the conversation unless you have two receivers.

One frequency in this band-segment, 156.8 MHz, is designated for distress and calling. The channel number is 16. Most marine communications originate on this frequency because radio-equipped boats and ships are required to listen on that channel when they're not engaged in communications. One ship calling another does it on channel 16. Once contact is established, they switch to some mutually agreed-upon channel for communications.

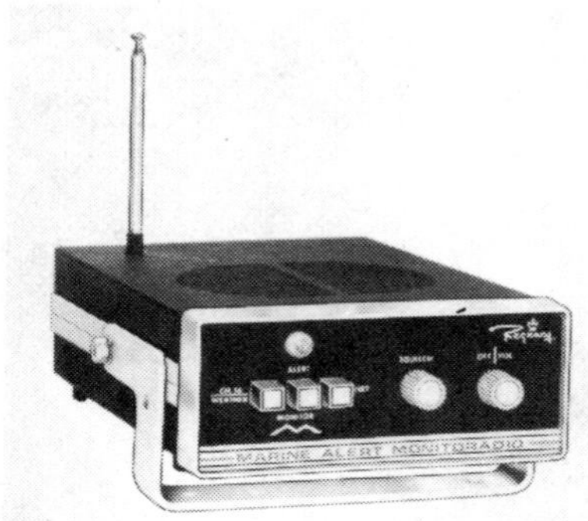

The operator of a neighboring yacht club or marina can tell you which marine channels are popular in your area. On rivers, channels 6 (156.3 MHz) and 26 (157.3 and 161.9 MHz—duplex) are popular. On tunable monitors, just swing the dial near 156 MHz; you'll hear whatever is on the air.

If you're an aircraft buff, you'll want a receiver that can pick up airplane communications. They take place between 118 and 136 MHz. That's vhf.

But there's one major difference between aircraft vhf and the vhf used by police and fire departments (in the 150-MHz band). Aircraft transmitters use amplitude modulation (a-m) instead of fm. Equipment to handle fm properly is more bulky than for a-m and airplanes need every spare ounce trimmed off. Too, because of their altitude, airplane transmitters reach fantastic distances with very little vhf power. The extra sensitivity and quietness of fm equipment is not necessary.

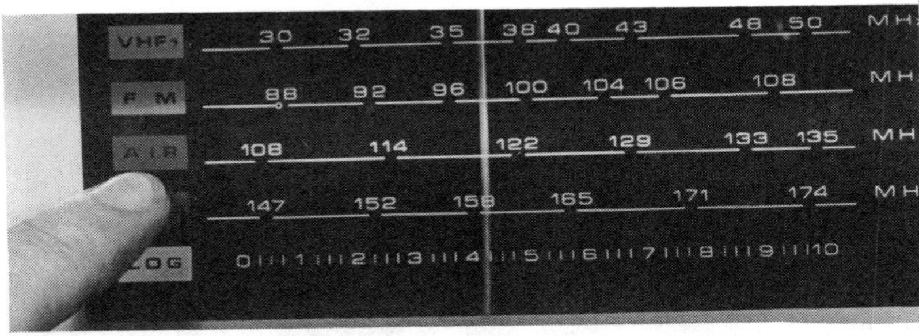

Aircraft navigation stations transmit between 108 and 118 MHz. So, the aircraft band actually extends all the way from 108 to 136 MHz. You can hear aviation weather on frequencies between 108 and 119 MHz, if you're near a ground station. From 119 MHz on up, you hear pilots and control tower operators, air traffic controllers, and the like.

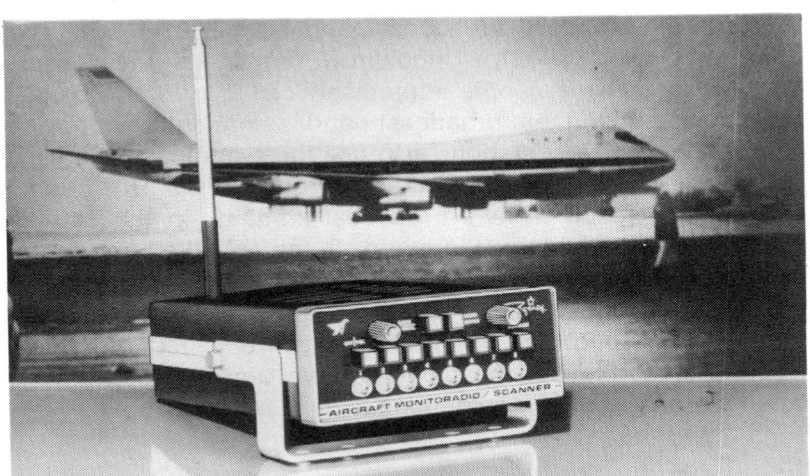

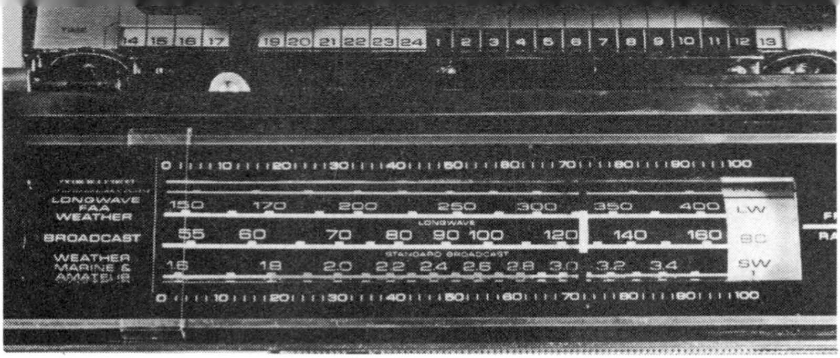

The fm broadcast band is not considered shortwave, even though its frequencies are vhf. They reach from 88 to 108 MHz. Many multiband radios nowadays include this band, so the receiver can be used for entertainment as well as shortwave listening.

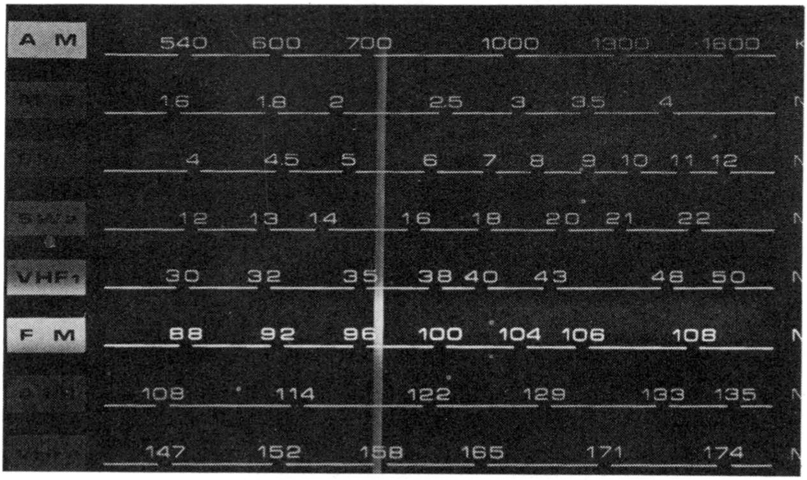

Nor is the standard broadcast band from 540 to 1600 kHz considered shortwave. It is medium wave, and its transmission mode is a-m. Some people imprecisely call it "the a-m band." Even calling it "the a-m broadcast band" is not entirely correct; shortwave broadcast stations also use the a-m mode. The most common misnomer is calling it "the broadcast (or BC) band."

There are medium-wave broadcast stations, in the 540- to 1600-kHz range, all over the world. Some so-called shortwave listeners are actually medium-wave buffs. They try for those faraway (and weak local) broadcast stations down in this part of the radio spectrum. These listeners are called *broadcast DXers.* Don't rule out the medium-wave a-m broadcast band when you're hunting for something to have fun listening to.

Chapter 4

Shortwave Listening Equipment

How much money you spend on shortwave radio depends on several factors. Will your listening be serious or merely casual? Do you plan to identify the stations you're hearing and try for QSLs from them? Will you listen mostly in the daytime or mostly at night? Is all your shortwave listening to be at home or do you expect to take your radio along on trips? Do you demand the best of everything or can you be satisfied with just adequate? Can you afford more than one receiver or should you buy the most multipurpose receiver you can find?

The pages that follow should help you make buying decisions.

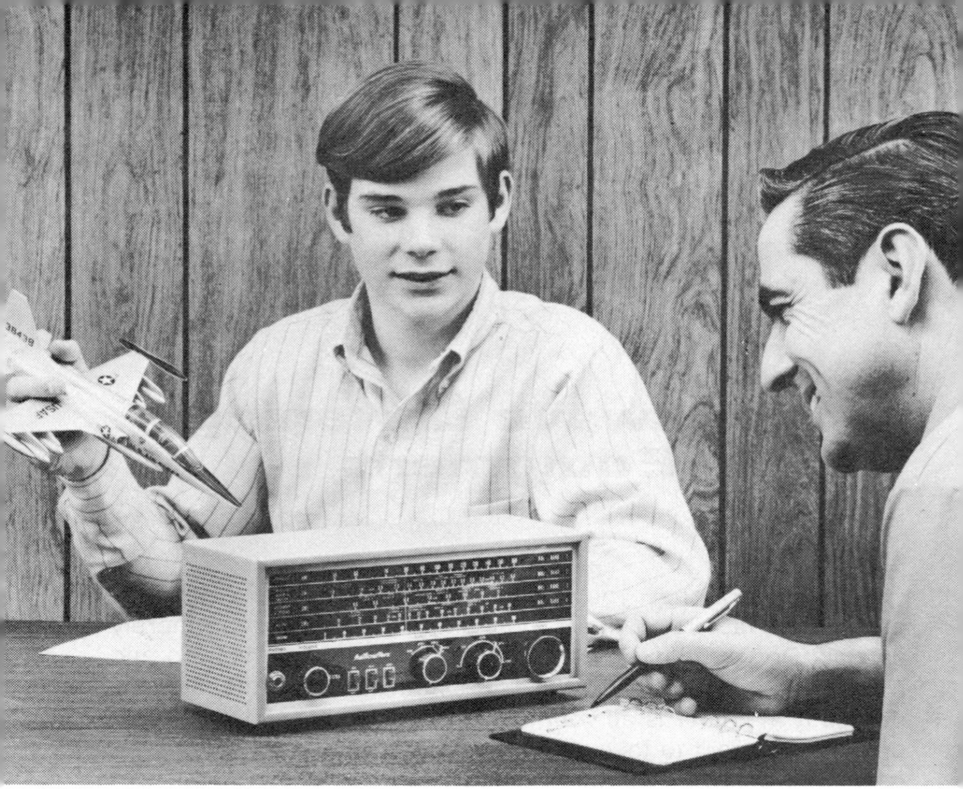

Using a carefully strung antenna, judicious choice of listening times and frequencies, and knowledgeable operating technique, a relatively inexpensive receiver can produce excellent long-distance reception. On a limited budget, consider the small general-coverage shortwave radio. "General coverage" means usually that the dial tunes from 540 kHz through 25 or 30 MHz —in several "bands," of course.

You can hope your little all-wave receiver (as general-coverage units are sometimes called) contains a bandspread knob. With the main tuning control set *near* the station you want, the bandspread control lets you "fine tune" closer than you can with the large dial. This is necessary when twenty megahertz of frequencies are squeezed into 6 or 8 inches of dial space; that's the case with higher frequencies in most inexpensive receivers.

The real factor, in using small shortwave sets, is the enthusiasm, patience, and experience you bring to the task of sorting out stations. Walls have been filled with QSLs earned with a low-cost shortwave radio and one long piece of wire.

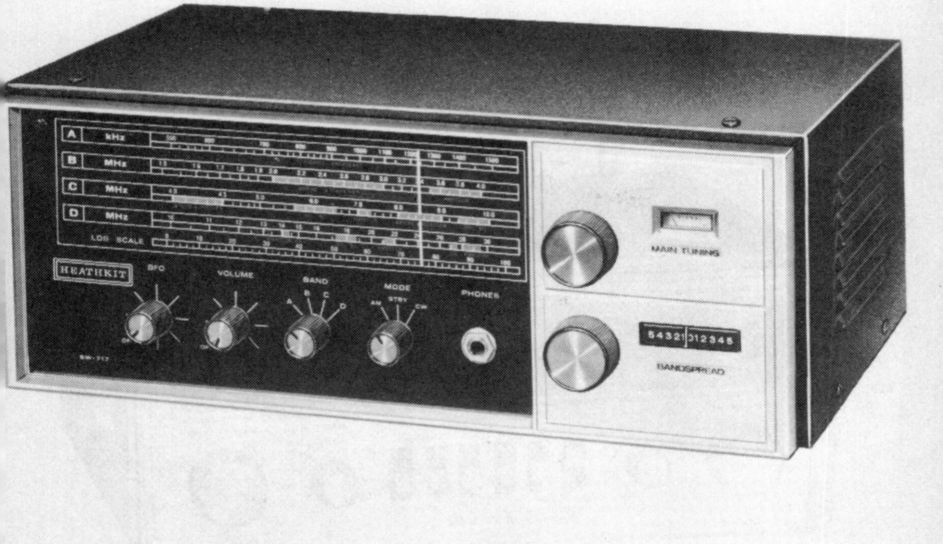

There's another way to save money when you start out as a shortwave listener. Build your own receiver!

Don't scoff. You can do it, if you can read simple instructions, hold a soldering gun, and turn a screwdriver. Hundreds of these units are built every year by people who don't know a microvolt from a tuning capacitor. And the receivers work.

This model follows the frequency pattern usual with inexpensive general-coverage receivers. Four bands cover the frequencies from 540 kHz to 30 MHz. Band A tunes from 540 to 1500 kHz, band B from 1.5 MHz (1500 kHz) to 4.0 MHz, band C from 4 to 10 MHz, and band D from 10 to 30 MHz. A bandspread dial helps with tuning, especially at the high frequencies.

This set has a bfo for cw reception. No inexpensive receiver can handle ssb signals directly. However, there's a method outlined on pages 126 and 127 that will work—providing you have patience and your receiver has an adjustable bfo.

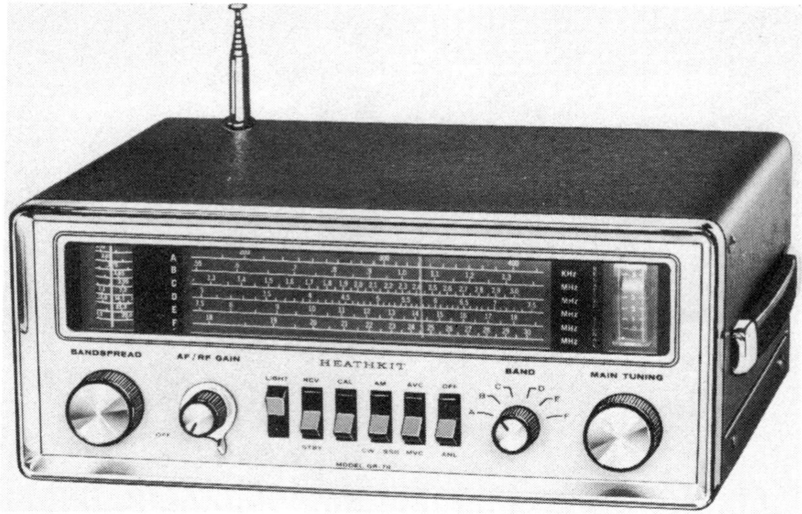

The most modern shortwave receivers are solid-state. That means the circuits use transistors instead of tubes. This cuts down heat and theoretically improves dependability. But another benefit is really more important. Solid-state amplifiers in the radio make less noise than amplifiers using tubes. Weak stations have a better chance of being heard.

The unit pictured here demonstrates features you can look for when you spend a little extra on a general-coverage receiver. The 540 kHz to 30 MHz occupies five bands rather than four. Hence, only twelve megahertz are spread over each dial-length at the high-frequency end. The crowded-at-night 1.8- to 3.0-MHz band covers two-thirds of its dial length, spreading those frequencies out. This set also offers the 190- to 410-kHz long-wave band and a bandspread calibrated for nine popular shortwave bands.

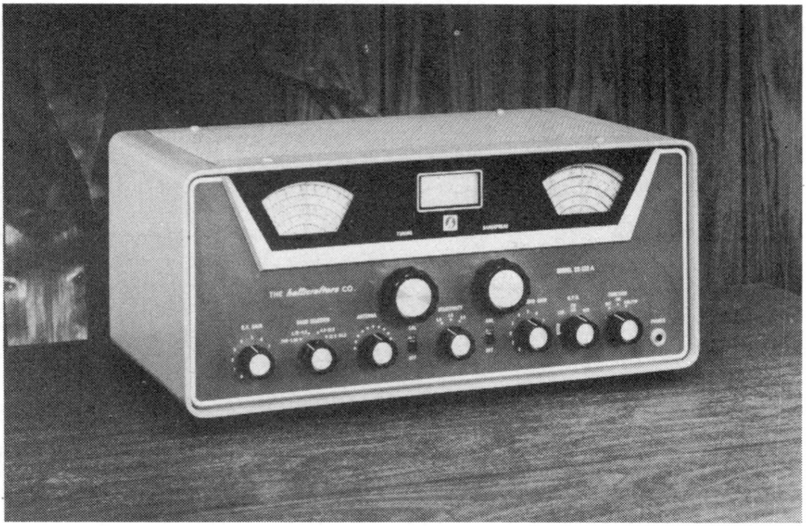

Spending a few more dollars adds more features to the receiver. But do they add to your listening? You have to evaluate.

Calibrated bandspread puts more stations at your fingertips. Yet, bandspread doesn't always separate those side-by-side foreign stations that come crashing in on a good reception night. *Selectivity* is the quality that lets you separate the tough ones from those around them. You can adjust the selectivity of this set narrow enough to just barely let a station's sidebands through.

The bfo knob is marked for the correct sides of a single-sideband signal. This isn't a true ssb receiver but it is geared to help you decipher the Donald-Duck chatter of ssb transmissions.

You could do without an S-meter (signal strength meter). If stations are closely packed on the dial, the S-meter means nothing. However, to report relative signal strength of a strong station, the meter gives you something to read.

Study features carefully before you let them convince you to buy a particular receiver. Be sure they actually apply to your use of the radio.

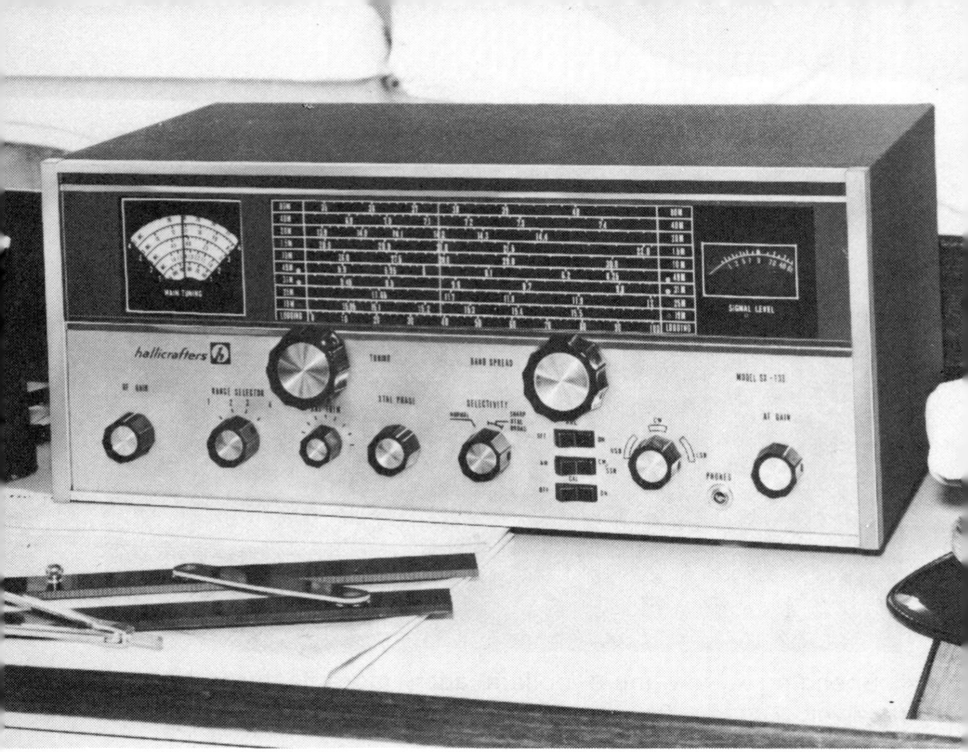

This solid-state receiver has a reversal of ordinary tuning arrangements, to a distinct advantage. Usually, a radio's large dial covers main tuning, and a small one to the side is for bandspread. Not so, here.

The large main dial spreads out six major international shortwave bands (13, 19, 25, 31, 41, and 49 meters) and five chief ham bands (10, 15, 20, 40, and 80 meters). To boot, the 10-meter ham dial tunes CB frequencies.

In effect, this becomes a specialty receiver for both ham and international shortwave reception, but designed into a general-coverage radio. The smaller main tuning dial lets you tune any frequency from 540 kHz to 31 MHz. Although bandspread-dial calibrations apply only to nine specific settings of the main tuning dial, a little intelligent extrapolation (see page 121) lets you spread almost any small group of frequencies over one of those 10-inch dials.

A crystal calibrator helps zero-in dial accuracy. The bfo knob (labeled USB-CW-LSB) has markings to guide you in finding sidebands. Crystal filtering narrows selectivity. You pay for these features, but they're definite aids to shortwave listening.

When you pay more than $200-$300 for a receiver, you're in the big leagues of shortwave listening. One of the more important features you'll buy is accurate, drift-free tuning. An expensive receiver includes tuning mechanism and circuits aimed solely at delivering the station you're after, or at least putting you on its precise frequency.

Once the radio is warmed up, which may take 15 to 30 minutes, you pinpoint its dial with a built-in calibration arrangement. Barring drafts and battering, the tuning then stays exactly on frequency. If you change to another band, recalibrating takes only a few moments.

Of course, that's not the only feature you buy. You can expect accurate reception of ssb signals. Noise should be very low. Sensitivity, very high. Selectivity should be narrow enough for picky tuning among close-packed stations.

Expensive receivers have one drawback. They are almost never general-coverage. You have to settle for whatever bands they cover. Choose the receiver carefully.

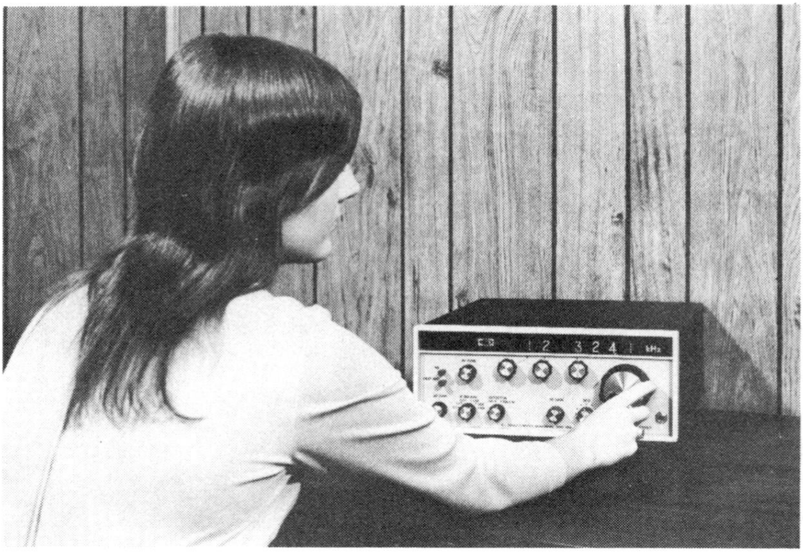

Here, too, you're looking at a high-priced receiver. It comes in one model for ham operators and another for the international shortwave bands. The key factor is that it's a precision receiver. Tuning is rock-solid, pinpoint accurate because of the digital readout, and internally calibrated.

But remember, it doesn't take a lot of money to be an accomplished shortwave listener. A patient student, with the little improvised gadget shown below and 120 feet of antenna wire, listens regularly to some of the world's top shortwave stations.

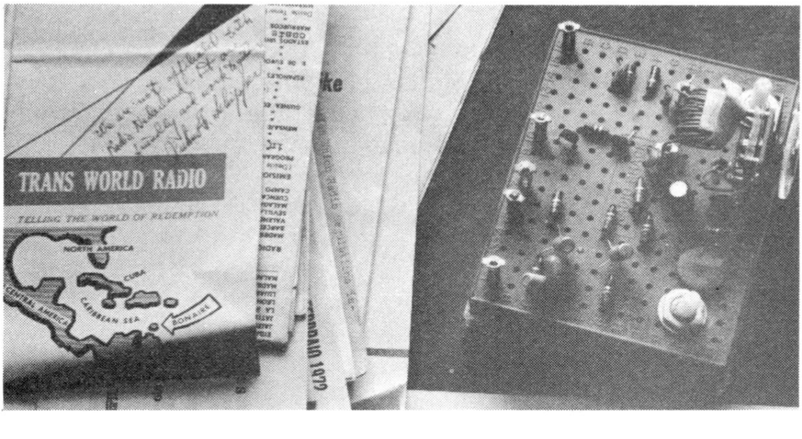

Here's what has been turning the younger generation to shortwave listening. It's simple and not too expensive. Best of all, you can take it to the beach, to the park, to ball games, or anywhere you go when you're a busy student. It plays on electricity (house current) or its rechargeable nickel-cadmium (Nicad) battery. When you're selecting a portable like this, study its frequency coverages carefully. Be very sure it doesn't skip over some band that's important to you.

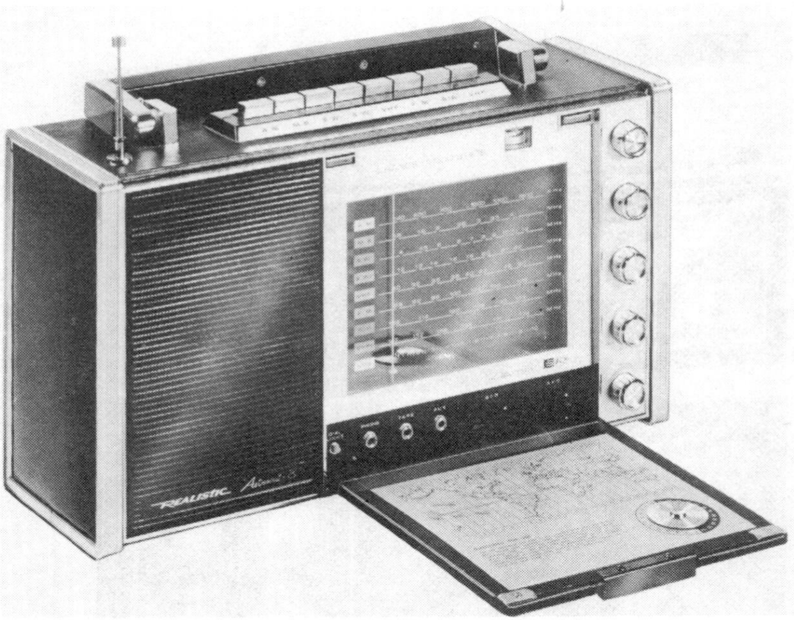

This is more elaborate, yet still entirely portable. This particular multiband set offers more coverage than most radios available today.

First of all, a-m and fm standard broadcast bands carry entertainment for you. Low-band and high-band vhf-fm bring you police, fire, boats, weather, and other land-mobile radio services. Three shortwave bands (one labeled Marine Band) encompass frequencies from 1.6 through 23 MHz. A bfo lets you sort out code stations—and maybe try for a single-sideband operator occasionally. To complete the coverage, a vhf a-m band from 108 to 136 MHz picks up signals to and from aircraft.

All these bands match the collapsible whip antenna built into the case. For extra sensitivity on the shortwave bands, you can attach the lead-in from a long-wire antenna. There's even a limited bandspread knob. For a medium-price portable, this model offers exceptional versatility.

Among long-time shortwave listeners, most agree that this is the granddaddy of all shortwave portables. The "Trans-Oceanic" (sold by Zenith) goes back quite a few years. It was the first high-sensitivity shortwave receiver to be made portable.

Of course, endless improvements have been added down through the years. Now the unit is all-transistorized, with an accompanying increase in battery life and reduction in noise figure. To the 1973 models, a weather band (161.0 to 164.0 MHz) has been added. (Continous National Weather Service broadcasts now emanate from transmitters in several U.S. cities on frequencies of 162.4, 162.45, and 162.55 MHz.)

The long whip antenna that is part of the Trans-Oceanic portable is all it needs for unbelievably sensitive operation on its shortwave bands. Besides standard a-m and fm broadcast bands, the Trans-Oceanic covers 200- to 400-kHz long wave, 1.6 to 9 MHz continuous in two bands, and the 13-, 16-, 19-, 25-, and 31-meter bands spread out for easier tuning.

Portables bring shortwave listening to outdoor living. At work or play, in office or park, the SWL with a portable radio can take the hobby along with him. With the versatility of today's super-portables, you take entertainment along too.

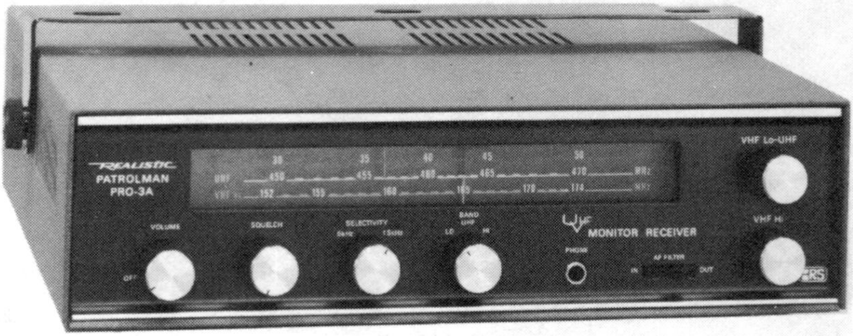

The public-services monitor has grown popular among listeners. Many nowadays don't even bother with shortwaves from overseas. There's enough excitement right around home.

Frequencies in the vhf spectrum don't travel very far. Essentially, they're line-of-sight. A certain "waterfall" effect operates for communication signals in the 30- to 50-MHz band. That is, these signals tend to spill over a hill that would block them and, to some extent, over the natural horizon. Thus, they seem to travel farther than higher vhf signals.

In a lesser degree, high-band vhf (150-175 MHz) signals experience the waterfall effect too, but communications in this band are considered dependable chiefly in line of sight. For uhf frequencies, transmission is strictly line of sight.

The key to distance, then, is antenna height. The higher the transmitting antenna—or the receiving antenna—the further the communications horizon. As a listener, you have control only over the receiving antenna. The higher you mount it, the more likely you are to hear mobiles across town as well as base stations. (The latter are situated to cover an entire city or county if possible.)

A recent development, the monitor scanner, simplifies listening to specific stations in the public service bands. Here's the way it works.

When you buy the unit, you decide which vhf and uhf stations you want to listen to regularly. The dealer should know the local frequencies. He'll install frequency-tuning crystals for the services you plan to monitor. Then, when the receiver is on and you're listening, circuits inside scan from one frequency to another. When a signal from one of the stations is picked up by the antenna, scanning stops at that "channel." You hear only the message on that frequency. When the message ends and the dispatcher quits holding down his transmitting button, the receiver resumes scanning.

You can thus listen for several stations, almost simultaneously, with only one receiver. The monitor "holds" only when a station actually comes on the air and only during that transmission. With a scanner that handles all three bands—low vhf, high vhf, and uhf—you can pick up city police, county police, state police, fire department, the local merchant police, and so on, even if one of these services uses more than one communications channel.

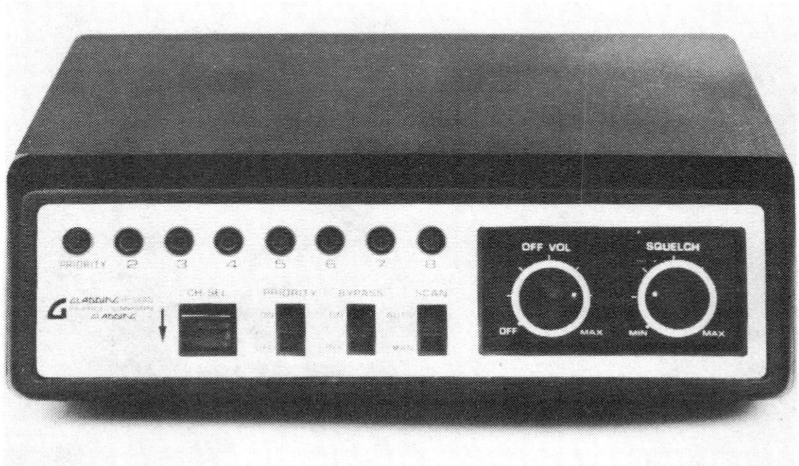

Certain companies produce scanning monitor receivers for special listening. The unit shown here, for example, scans the marine vhf channels.

You buy it with crystals in it for the eight marine channels most popular in your area. The first position is marked "Priority" and is reserved for channel 16. Other channels might be vhf WX (weather), channel 6, channel 68, channel 41, and so on. You could even reserve the last two positions for the ship/shore telephone channel in your locality (channel 26 is the most popular) and thus hear both sides of conversations on that channel.

As with any scanner, the advantage is that you don't miss messages on any channel your monitor tunes. If marine message traffic is important to you for any reason, the marine scanner assures you of hearing it. Most listeners, however, listen just for the excitement.

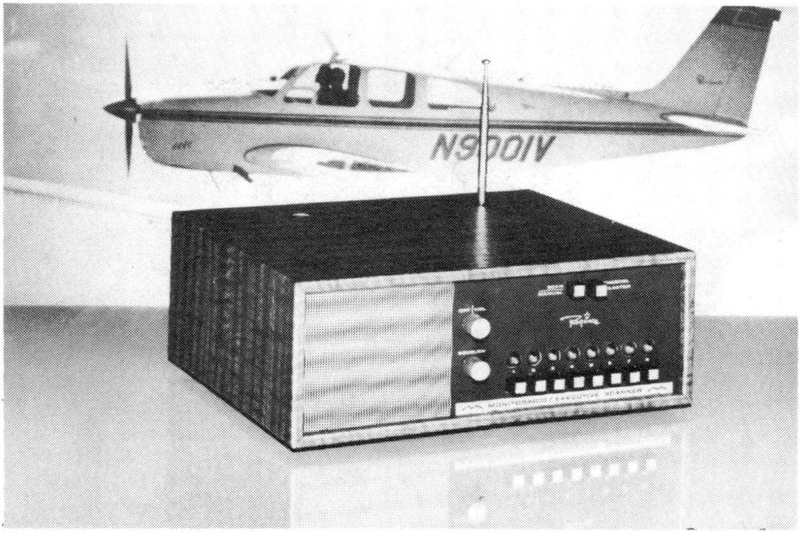

A scanner for the aircraft band gives you the same kind of versatility. You might, for example, have your aircraft monitor set up to tune the FAA control towers, ground control, the approach and/or departure control, and Flight Service station at your local airport(s). Perhaps your monitor will tune some private frequency used by a local aircraft company.

Your dealer should know what frequencies are used locally. If not, you can get them by calling the Federal Aviation Administration number in your telephone book. Even some small airports have transmitters for talking with aircraft; they're called Unicom transmitters. Phone the airport operator nearest you for his frequency.

Some monitors have built-in antennas. Others require that you connect an external vhf antenna. Remember, the higher your receiving antenna, the farther away you can pick up stations. For airborne planes, this doesn't apply; their altitude gives ample height.

One interesting factor about shortwave receivers is that most table-model or expensive shortwave sets come without any speaker. You have to buy a separate speaker. This isn't true of monitors or of portable radios. Your dealer usually has speakers to match new rigs. If you buy a used receiver, the matching speaker may not be available.

Some speakers are designed only for communications. Their cones operate best at voice frequencies and virtually ignore audio frequencies that contribute to hiss, static noise, and other unwanted sounds. They sound great for voice communications but tend to be muffled with standard broadcast sounds (music, etc.).

Any good speaker will do, but mount it in a box or baffle. Where you place the speaker matters little; beside the radio, on top of it, even under it if the size permits—all are okay. You can switch one speaker among several receivers if you need to. (More about this on page 92.)

You may do the majority of your shortwave listening using a headset. This has advantages. By blocking out room-noise distractions, you can concentrate specifically on what you're listening for. This can prove important when you dig down into a "mish-mash of chatter" to separate that one important signal you've hunted for a month or more. For high-speed code copying, a headset becomes almost a necessity because of the intense concentration needed. Too, you can hear voices and code better with a good set of communications (not hi-fi) headphones.

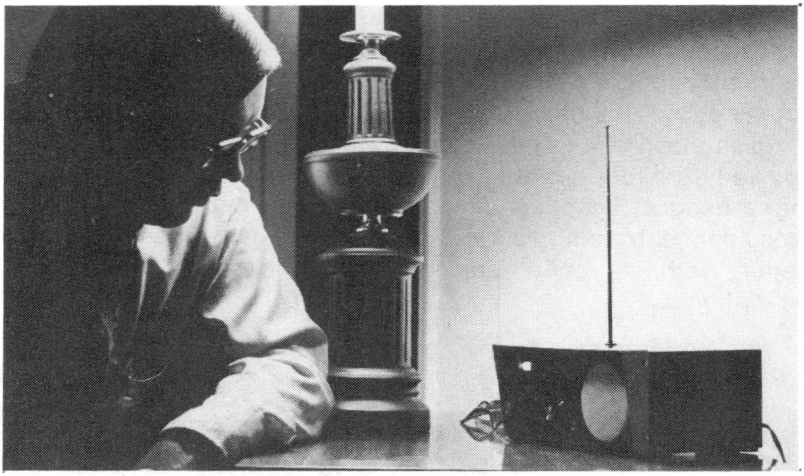

Here's a monitor buff who does something special with a headset. He listens mainly at night, after dinner. The kids make noise. The chatter of the receiver bothers his wife. What's more, he likes to listen to more than one radio at a time. So . . . he put together a device that lets him utilize his stereo headphones. (Yes, hi-fi, even though they're not the best for ordinary shortwave listening).

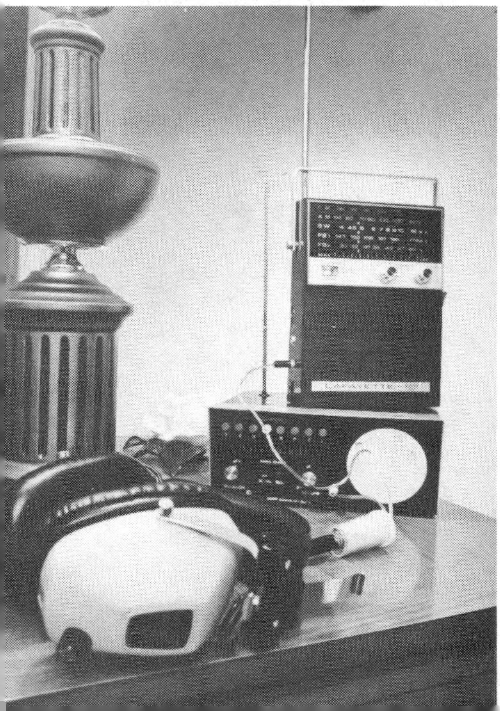

One small plug goes into the earphone jack of his police-band portable. Another small plug goes into the phone jack on his scanning monitor. The two-wire pairs from these plugs go to a *stereo* headphone jack he mounted in a plastic pill bottle. The jack is wired so the output of one receiver goes to the left cup of his stereo headset, and the other output goes to the right cup. He hears message traffic from the scanner with one ear and from the portable with the other.

The store where you buy shortwave listening gear has a stock of other accessories to make your hobby easier or more interesting. Headsets, special speakers, lights that make tuning easier, and electronic equipment to assure precise station-finding are available for a price.

One of the first "extra" investments you'll probably make, however, is in a 24-hour clock. It takes effort, even after you've used Greenwich time for awhile, to keep converting what you read on your watch to GMT. The serious SWL buys a 24-hour clock and keeps it set to GMT. He then knows what time it is anywhere in the shortwave world, merely by looking at his clock.

And where do you find your equipment? Look in the Yellow Pages of your telephone book under "Radio Communication Equipment & Systems." Also check under "Electronic Equipment & Supplies—Wholesale." The companies listed either carry SWL equipment or know who does. That will start you off.

A store specializing in communications equipment usually offers the widest variety of shortwave gear. Too, you can expect their personnel to have considerable know-how in communications electronics. Very likely, some of them are shortwave listeners or hams.

Ask your radio or tv repair technician if he knows who in town carries shortwave equipment. He might be able to recommend a company and even give you some pointers about shortwave receivers in general. The organization from whom he buys replacement parts may also sell shortwave radios.

Local electronic stores usually stock shortwave receivers. What's more, some of them know other shortwave listeners and ham operators you can talk to about equipment. Try to deal with businesses that obviously cooperate with shortwave buffs. You can depend on them for advice regarding antennas, technique, accessories, and many of the little things that make listening more fun.

Visit several stores. You may find one that specializes in communications. The owner/manager of the chain store pictured here has built a large clientele of ham operators, CB radio owners, and shortwave listeners. It's obvious, from this special communications center, that he knows the business and stocks the equipment you'll be looking for. What he doesn't have in the store he can order from the parent company.

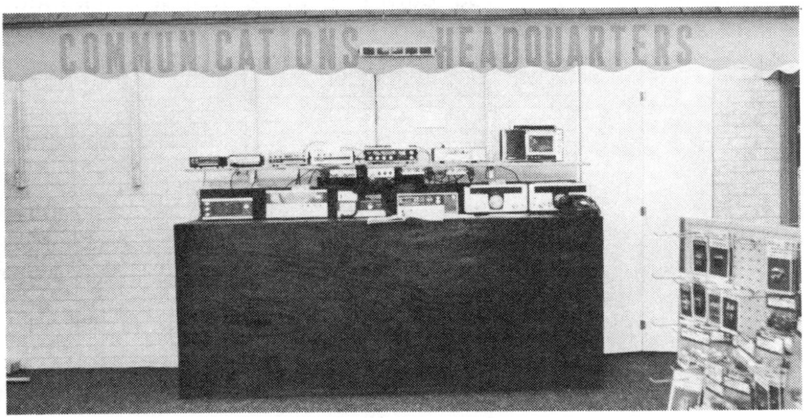

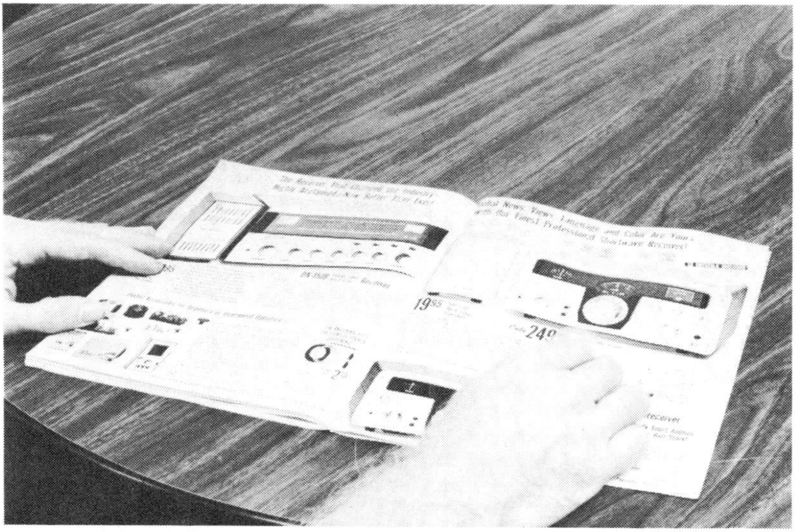

Many companies sell by mail order. Look for their ads in electronics, CB, shortwave, and do-it-yourself magazines. A postcard will usually bring their catalog. Study the catalogs. Read about various brands in detail, comparing features and prices.

Many of these companies also have sales outlets in your town, where you can see the merchandise. If you can find others who share your SWL interest and have some experience, ask them what brands seem to do the best job. Then make the best choice you can. Chances are, once you've gone through the rest of this book and applied what you learn to your selecting, you'll make a satisfactory purchase.

Chapter 5

Setting Up Your Listening Station

Where you stack your shortwave listening equipment is not important, except to whoever has to live in the same house with you and it. If you have plenty of room, the SWL "shack" should go somewhere out of the way—where you have privacy and where the sudden squawk of a radioteletype station won't offend anyone else's ears.

When space is limited, ingenuity and patience are the chief ingredients of a successful SWL arrangement. Careful thought and inventiveness can make even a kitchen corner comfortable to use. Note, for example, the "tuning board" here, resting on an open cabinet door. It makes up for the limited width of the cabinet top. This SWL can turn that tuning knob for hours without arm fatigue.

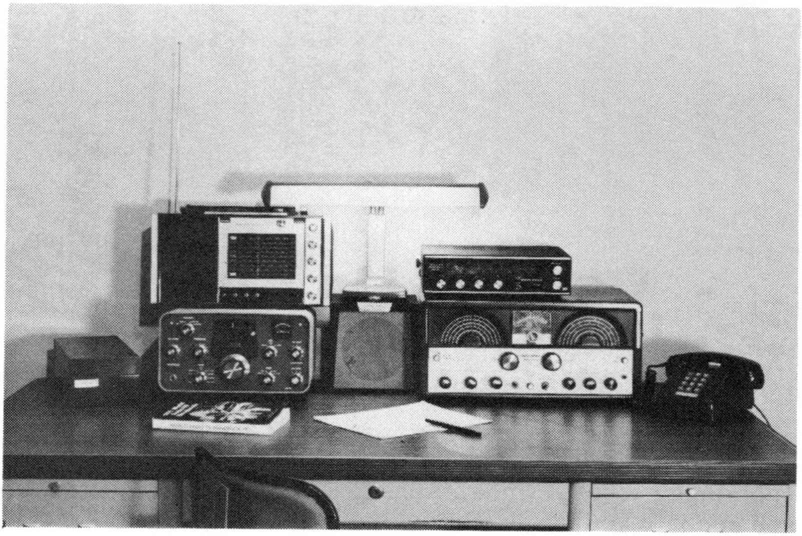

You might be fortunate enough to afford an elaborate "all-bands" setup like this one. A spacious desk, receivers for just about any kind of listening you can imagine, and plenty of light—what more could you ask? A box holds your card file where you keep track of stations you've logged and those you haven't been able to find yet. Most of all, the neatness of this station is a pleasure to you and your visitors. (Of course, some people don't mind clutter.)

Don't get the idea that you need a fancy setup right away. Learn to use one piece of equipment at a time. The shack pictured here was assembled over several years. The general-coverage receiver is an older model, but sensitive. The special receiver at left is to go after those almost unreachable stations every SWL prizes. The multiband portable adds two or three bands not included in other sets and is available to take along on trips. The three-band monitor receiver is brand-new; it became necessary when local police and fire departments moved to the uhf band.

There's duplication here, but that's not necessarily bad. It permits some simultaneous "fishing" when one kind of listening seems a bit slow, yet you don't want to miss something.

This is another orderly listening post. It is elaborate in its own way. The world map, GMT clock, and shortwave schedules suggest activity and above-average interest. The QSL cards surrounding the map prove beyond a doubt that one good receiver is all any SWL needs. Using your shack regularly and intelligently brings in QSL cards you can be proud of.

This station is set up for comfort. The radio is in easy reach. A clipboard holds sheets for logging, and the schedule book is right at hand. The clock is easy to see, the map quick to study. Ashtray, tape, pencils, everything close by: that's a key to hours of SWLing without fatigue.

An easy-feeling chair makes long listening hours bearable. An uncomfortable backbreaker may spoil your enthusiasm.

The simple kind of office desk chair, with arms, offers a lot in the way of comfort. Cloth coverings for the seat and back add even more. If you're shopping, hold out for the kind of chair that has the back, arms, and seat connected all in one piece. The break-away-back type lets you lean back but in the most awkward of positions. With solid seat/back, you can lean back and stay comfortable, yet remain in an operating position.

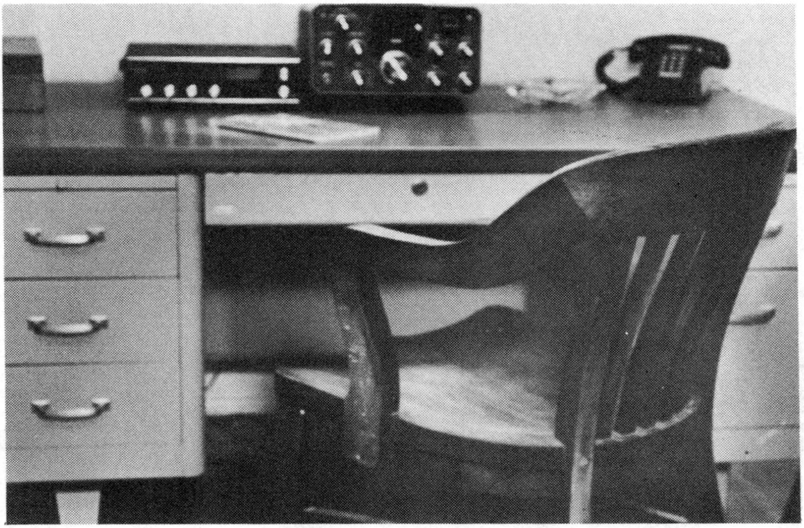

Another kind of chair that is comfortable for shortwave listening is the "captain's chair." These are scarce. If you live around a small town, you may find the bank has some stored away. Otherwise, haunt the second-hand stores in your city to find one. Again, a cushion in the seat, preferably with a cloth covering, adds measurably to your hours of comfort.

Don't forget a couple of extra chairs for company. You may find your shortwave listening station a neighborhood attraction. Teen-agers particularly find an hour or two hanging around electronic communications gear fascinating. (You may even stimulate some new SWL stations.) Used dinette chairs make good and inexpensive visitor chairs. Decide how frequently you want people in your listening shack; if not often, your extra chairs should not be too comfortable.

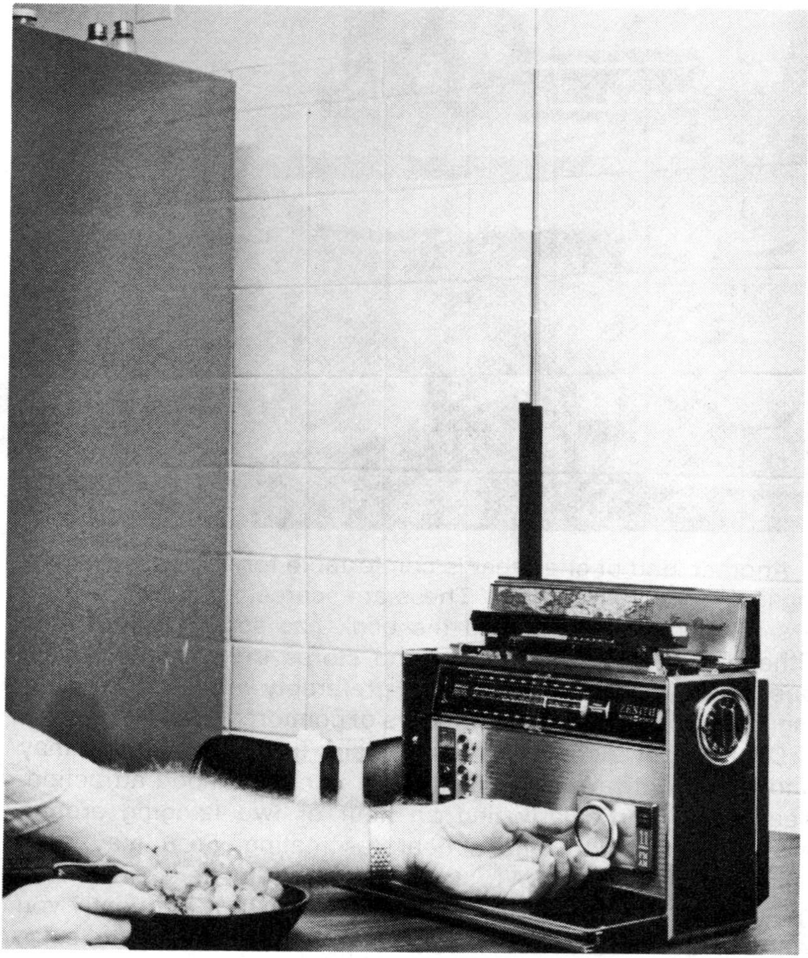

Maybe you just don't have any place in the house to set up a formal listening spot. You can listen anyway. The kitchen table is just the right height and some dinette chairs are mighty comfortable.

If you have a radio with an efficient aerial of its own, you've got it made. You can also run an antenna lead-in through a nearby window or up from the basement behind the refrigerator. When you're not listening, just coil the wire up out of the way and put your receiver in a closet.

This setup is thrown in just to make you envious. The fellow owns a pickup-truck camper. For summer vacations, it goes with the family. But the rest of the time, he has his own shortwave listening shack—literally. A couple of antennas strung up in nearby trees, a table inside converted to a bench, and he has the smoothest SWL shack in the neighborhood. Nobody annoys him and his listening bothers no one. Nice setup, if you've already got the camper. Costly, otherwise.

Some small details about your shack are important. Like the kind of electrical outlets you use. This is called a three-wire outlet. The round hole takes the grounding prong of a three-wire power cord. Modern receivers, particularly those with metal cabinets, come with this kind of plug. The object is safety.

Don't use an adapter to convert a three-wire plug to fit a two-wire outlet. Instead, have an electrician install the right kind of outlet. The cost is small. But *be sure of this electrical ground.* Your life might depend on it.

This isn't the same as the antenna ground system. Furthermore, the electrical-system ground *must not* be used as a ground for the receiving antenna system. (Nor vice versa.) Page 107 describes a proper antenna ground.

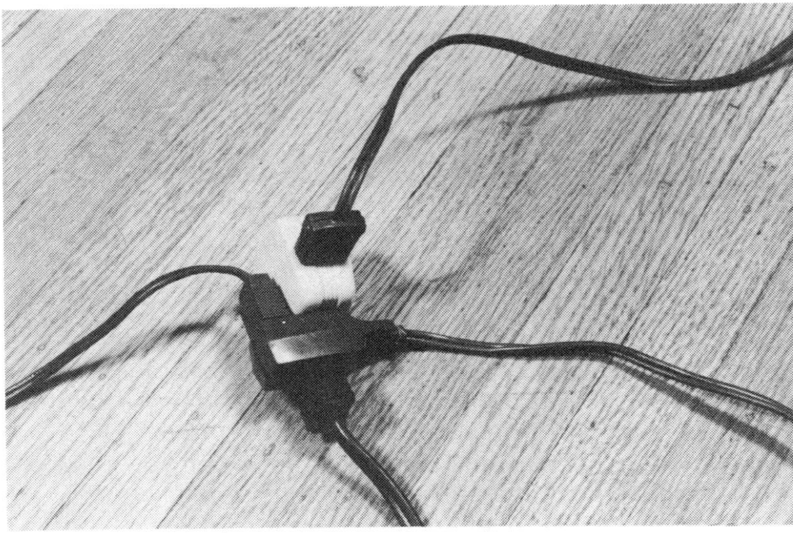

Avoid this dangerous mess. Put in enough properly connected three-wire receptacles to take care of all your receivers and accessories. Don't forget enough outlets for the lamps and clock. The old cube-tap octopus may louse up your reception some hot night and let you hear the fire-department calls first-hand—right off the trucks.

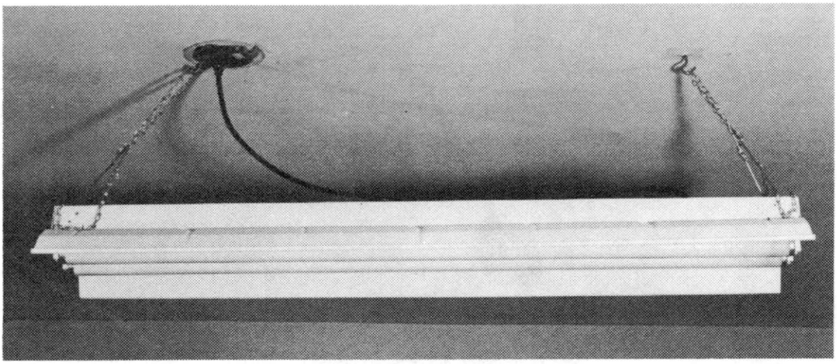

You'll need plenty of light. A 48-inch twin-bulb fluorescent fixture sheds excellent general light for your shack. But it's electrically noisy. Put it on a switch and make sure the switch is near your listening position. There are times when you'll prefer to work from the light of the dial lamps in the radio.

But arrange for full-room lighting in your shack. You'll drop pencils, little screwdrivers, and a dozen other things. Without regular room lighting, you'd better keep a good flashlight handy and have plenty of fresh batteries.

Probably, you'll use general lighting for logging, too. And to look up schedules, hunt cards in your file, or show visitors your latest QSLs. For all these, fluorescent lighting beats incandescent bulbs all hollow.

When you're working out your lighting scheme, stop to consider what movements you'll be making. You'd not want a lighting setup like the one above, no matter how attractive it might be. For logging and for some listening, the single lamp might be okay, but you can't see to tune the radio.

Nor do you want general lighting for tuning. It will do, but it's not the best way.

If you're right-handed, install a small 100-watt bell lamp that shines directly on the front of your receiver from over your left shoulder. The angle shown below works out well. If you tune with your left hand, situate the lamp to shine over your right shoulder.

Most important of all, put the switch for this lamp somewhere in front of you, preferably near the receivers. You'll thank yourself a thousand times for this special tuning light and for not having to turn around and reach to switch it on.

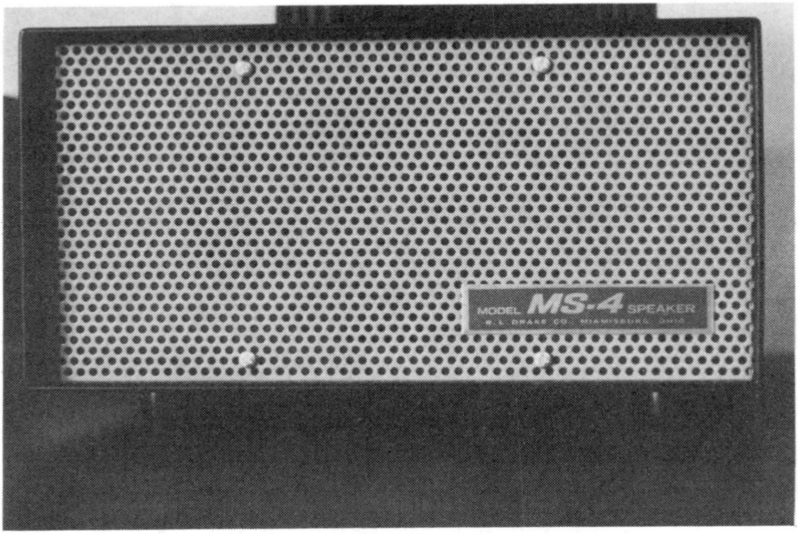

While you're wiring up a switching panel, give some consideration to using only one speaker and switching it among the several receivers. That precludes using more than one receiver at a time, but you may do that seldom anyway. Or, use two speakers and fix the switching so you can connect any receiver to either speaker.

With transistorized receivers, be careful in making this kind of setup. Be sure each solid-state radio has a load resistor connected across the output when no speaker is connected. Otherwise, you'll destroy the output transistors by running the receiver without a speaker. The best modern receivers have this protection, but check to make sure.

Consider the same kind of hookup for your headset. One jack and a switching system obviates pulling the plug and moving it to another receiver every time you change your mind about what to listen for.

You might even copy the trick on page 76, but in a larger way. Use a stereo headset and a stereo jack. Then wire the switches so you can connect either earpiece to any receiver you choose. The wiring would be the same as for two speakers (facing page).

If you are not up to such complicated wiring, hire someone who is. Tell him what you want and he'll put it together in a jiffy. It's relatively simple. Your shortwave equipment dealer might draw you a diagram on how to wire it, or even—for a price—have a special panel made up for you.

The prime objective is convenience of operation. The few dollars are well spent if you plan to spend many hours in your listening shack.

Speaking of convenience, another minor detail often gets overlooked. The knobs and switches of many shortwave receivers are arranged along the bottom of the front panel. This puts your hand down against the desk or table when you're trying to use the knobs. Your wrist can get mighty cramped after awhile.

Prop your equipment so it's tilted upward. Longer front "feet" are the slickest way. You can glue an extra length onto each front foot. The receiver in the photo comes already tilted. Not only does this make the knobs easier to handle, but you can see panel markings much better than when the front panel is straight up-and-down.

Watch for every possible alteration you can make to improve your comfort as you operate your listening station. Only if you can work at ease will you spend the time for real listening satisfaction—no matter how expensive your equipment.

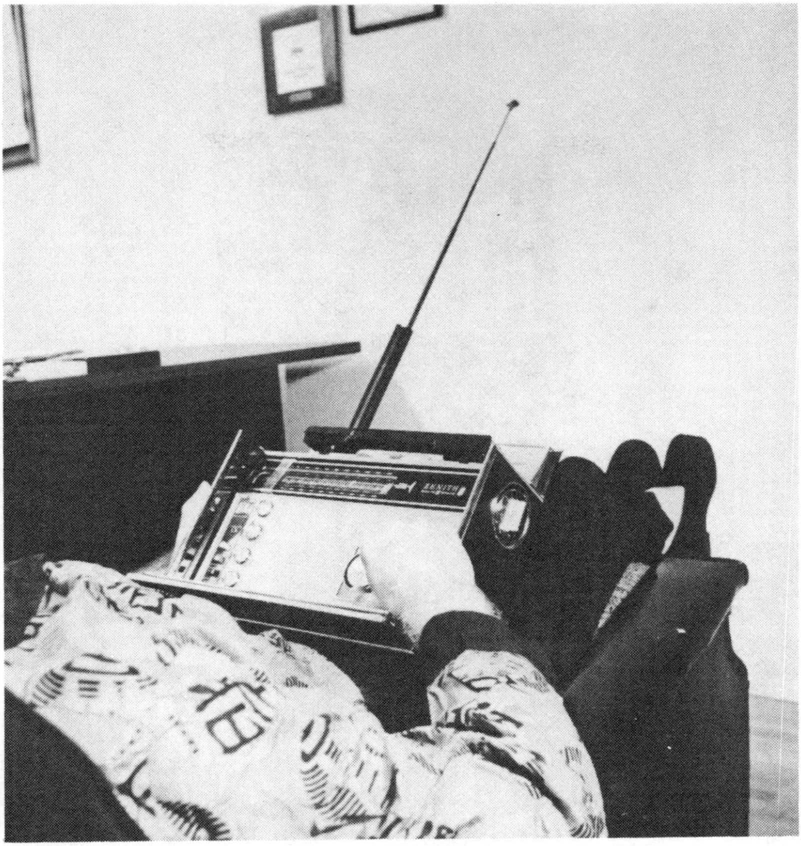

Maybe this is a little *too* comfortable. But, on the other hand, why not? If the excitement of shortwave listening can't keep you from drowsing, maybe you need the nap anyway. You'll probably awaken refreshed and finally tune in that weak pirate station down in the South Seas—the new one nobody in your area has been able to hear. Or did you just dream that . . . ?

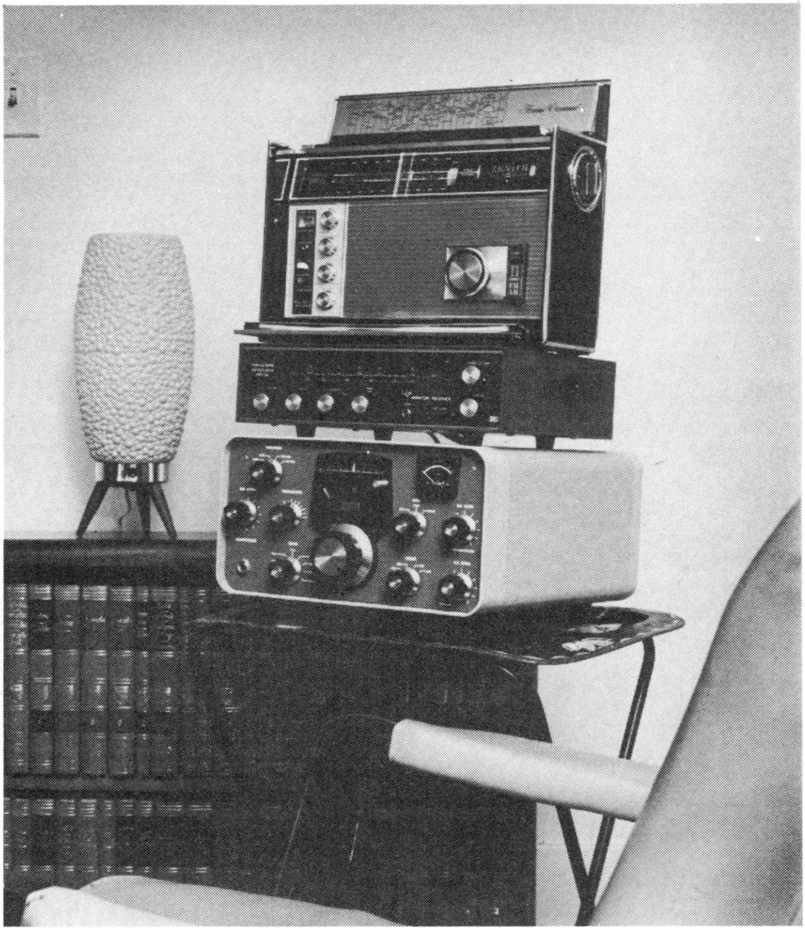

The chair looks comfortable, even if the equipment positions don't. And that TV tray appears a mite rickety. The whole shebang may end up on the floor.

Reaching up or down to tune your shortwave radio can be very tiring. You'll probably not stick with the hunting process more than a few minutes. And you won't be drawing in those tough QSLs.

The point is that you can listen to shortwave just about anywhere you want to set the equipment up. All you need is electricity and an antenna.

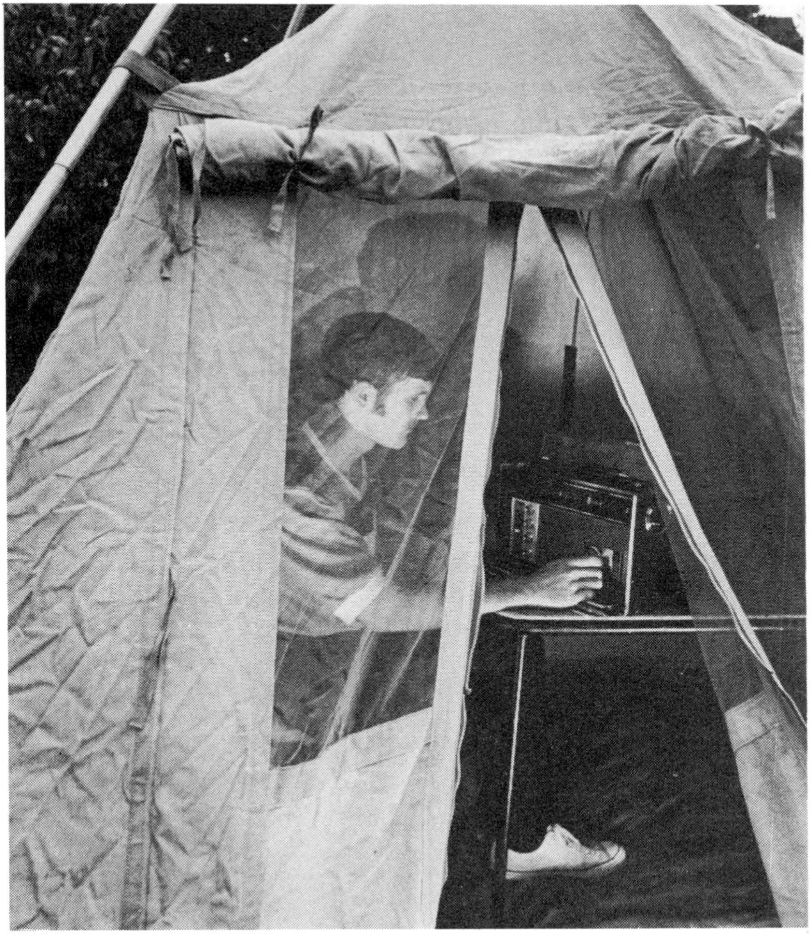

And maybe not even the electricity. This battery-operated listening shack has traveled over most of the United States and some of Canada. That's one way to grab some of those hard-to-get stations—move closer to where they are. (You can't get that Alaskan standard broadcast station? Try from the Yukon. Hawaii? Try from Mt. Whitney.)

All that this really proves is that you can set up a shortwave listening facility anywhere you take a notion.

Yes . . . anywhere. And you may be surprised at the interest you create. The most unexpected people become fascinated with this hobby, in one way or another.

Chapter 6

Antennas for Efficient Reception

This isn't the world's best way to hook up a shortwave receiver; yet a bedspring has been known to work—and do a fairly competent job. Two qualities make it a suitable antenna for DX reception: (1) It's metal, and (2) its wires present a lot of surface area.

The metal is essential. But other criteria mean more than the second factor. As frequency goes higher, these criteria assume greater importance to the shortwave antennas. This chapter explains a few of these critical qualifications.

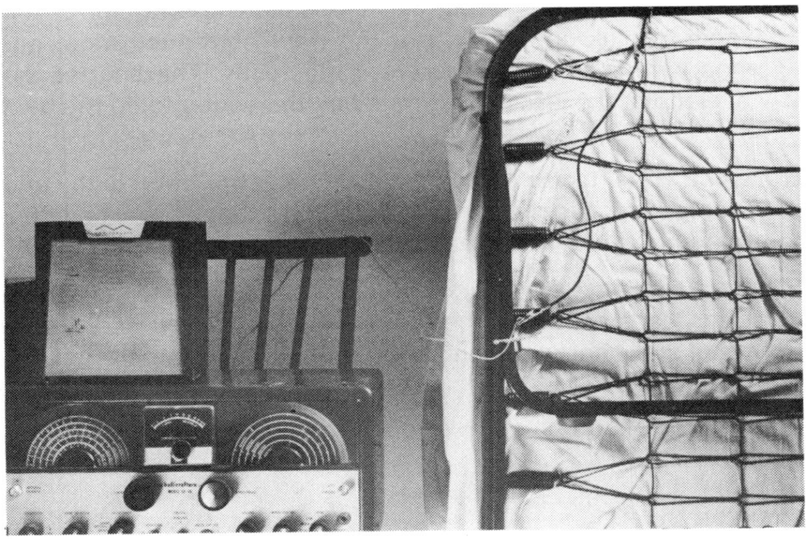

If you want to pick up radio signals from exotic places like this, you'll need an efficient receiving antenna. Three factors combine to make any given antenna poor, good, or excellent. These are: length, height, and directionality.

Length: This relates directly to the wavelength of signal you expect to pick up. The longer the wavelength, the more antenna wire required to receive it efficiently. The shorter the wavelength, the less the antenna length needed to do the best job.

Height: This factor applies mainly to high, very high, and ultrahigh frequencies. It would apply to longer wavelengths, but height enough for real efficiency would be prohibitive.

Directionality: A wire or a rod picks up radio signals best from broadside, unless certain connections alter the pattern. Directional shortwave antennas utilize this characteristic.

Antenna *length* should fit the wavelength but the size of a yard won't always permit that. An alternative is to "tune" electronically whatever wire length you can manage. That's done either with a *loading coil* or with a *trimmer capacitor.* The latter is the simplest and many shortwave receivers have an adjustable capacitor built right in.

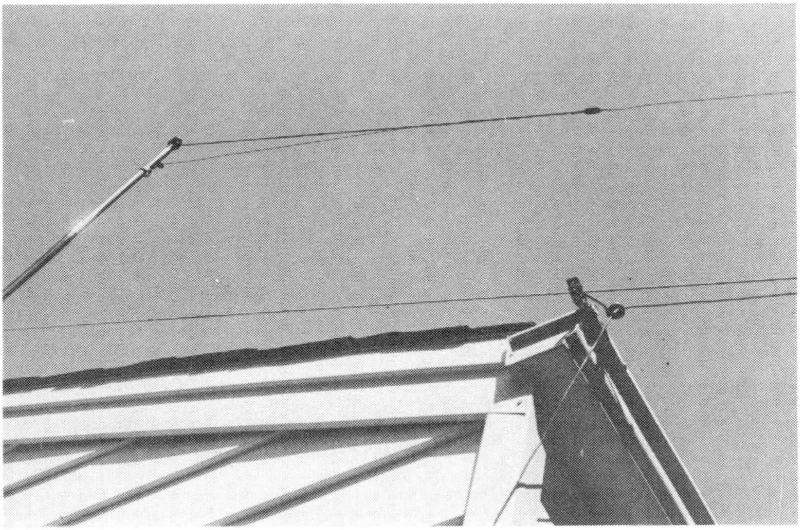

Up to 25 MHz, a half-wave antenna proves effective. It can be a single long wire of one-half wavelength, with a lead-in wire at the center. From 10 MHz on up, a doublet—that is, two wires separated at the center—does better. This takes a double lead-in. A half-wave dipole at 21 MHz figures out to about 24 feet long—12 feet per dipole.

Above 25 MHz, quarter-wave and five-eighths-wavelength antennas work well. The antenna can be formed of a single rod, usually held vertically. A quarter-wave "whip" for receiving 50 MHz figures out to almost 5 feet long. The length at 150 MHz is about 18 inches; and at 460 MHz, between 5 and 6 inches.

The heights of antennas for frequencies below 25 MHz go by the rule-of-thumb: the higher the better. To get away from an earth phenomenon called *image effect,* an antenna should be above the ground at least three wavelengths. A 25-MHz longwire would need to be nearly 120 feet high. This is not practical at all and is worse at lower frequencies.

But a quarter-wave whip antenna can be altered in such a way that the earth has little effect. The manufacturer adds a set of rods called a *ground plane* at the bottom. That isolates the antenna from image effect.

Yet, height remains important at vhf and uhf. That's because these signals stay near the ground and travel in a straight line. The higher the antenna, the farther the *radio horizon.* So, you mount vhf and uhf antennas on high towers.

Antenna directionality arises from the nature of radio waves. They travel transversely. A horizontal half-wave long-wire antenna picks up signals best from either broadside direction, *if* the lead-in is at the center. Connecting the lead-in wire at one end alters the pickup pattern, because of *end effects.* The antenna then picks up best from endwise.

However, frequency gets all tangled up in these generalities. A long-wire half-wave antenna at 20 MHz is full-wave at 10 MHz, two wavelengths at 5 MHz, and so on. The pickup pattern of this wire for those frequencies shifts in a way that reduces directionality.

Above 25 MHz, rod and whip antennas are practical because of their size. Rod antennas usually are doublets—called *dipoles.* They ordinarily lie horizontally and pick up best from broadside. Add more than one dipole, in an array called a *beam,* with some dipoles grounded instead of connected to the twin-wire lead-in, and you add extreme directionality in the broadside direction opposite the grounded elements. Rods of a beam can be vertical, too; directionality works the same. A rotator can point the *lobe* of greatest pickup anywhere you wish.

A whip antenna is omnidirectional—doesn't favor any one direction.

Selecting a long-wire antenna for shortwave listening needn't be as complicated as the preceding pages make it sound. Check the instruction manual that came with your receiver. It may give instructions for the one best antenna length to match the input circuits of your set. One receiver, for example, works beautifully with a straight 50-foot antenna wire. Try the manufacturer's idea first.

One young shortwave listener solves his antenna problem with a 75-foot length of heavily insulated wire. He stretched it between the house and garage, a mere 7 feet off the ground. A wooden clothes prop keeps it from sagging, and his mother hangs the wash on it. He hasn't reported what kind of clothes pull the best DX. Wet, probably.

It takes a long, long piece of wire for the medium-wave broadcast band. A half-wave at 1000 kHz, at band center, is more than 450 feet long. The problem takes a drastic solution.

Winding the wire into a coil reduces the length of wire needed. Adding a ferrite core reduces it more and multiplies efficiency manyfold. That's the antenna in any little a-m radio.

The broadcast-band DXer can wind 120 feet of wire on a box 1 foot square, with a 365-pF capacitor across the coil for tuning. That connects to a regular shortwave-type receiver.

A special broadcast-band DX (BCDX) antenna called the Spacemagnet boosts the strength of signals from one direction. It nulls out unwanted stations that are on the same frequency but in some other direction.

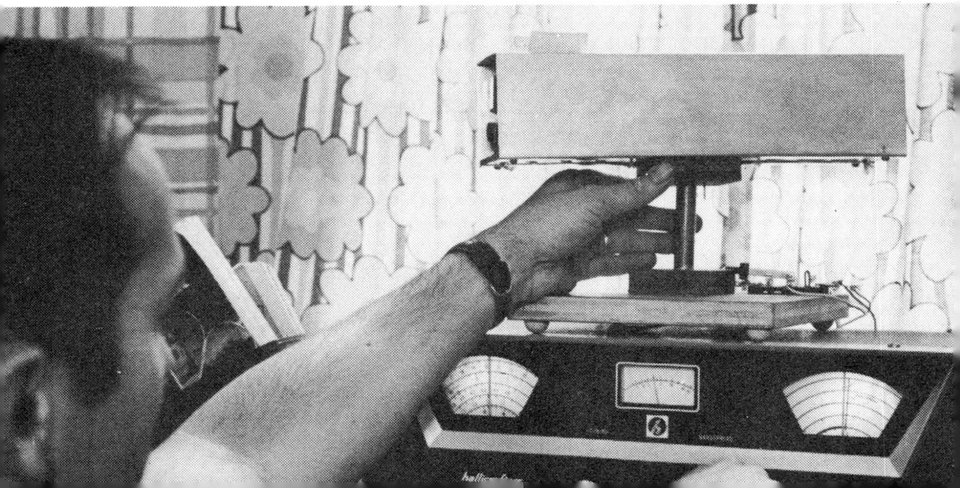

In case you're wondering how to put together a shortwave doublet, here's a closeup of how to keep the wires apart electrically at the center. Use an insulator. Also you can see the way to attach the double lead-in wires. Then twist the two lead-in wires together all the way to the receiver.

You support the whole doublet antenna with two more insulators (one at each end). TV-type standoffs support the lead-in wire as you bring it down the side of the house. Put the center of the antenna near the window through which you plan to bring the lead-in. Or, drill a lead-in hole at a point of entry near your shortwave listening position.

At the receiver, the doublet antenna connects to the two terminals marked "A" or "ANT." A coupling link between one "A" terminal and ground must be removed. The link is there for when you connect up a single-wire antenna lead-in. The "ground" or "G" terminal goes to a special radio ground rod (see facing page), no matter which type of antenna you're using.

The station grounding system has two individual facets. One ground goes through the electrical system, using three-wire wall outlets as described on page 88. The antenna system must be grounded too, *and that must be a distinctly separate ground.*

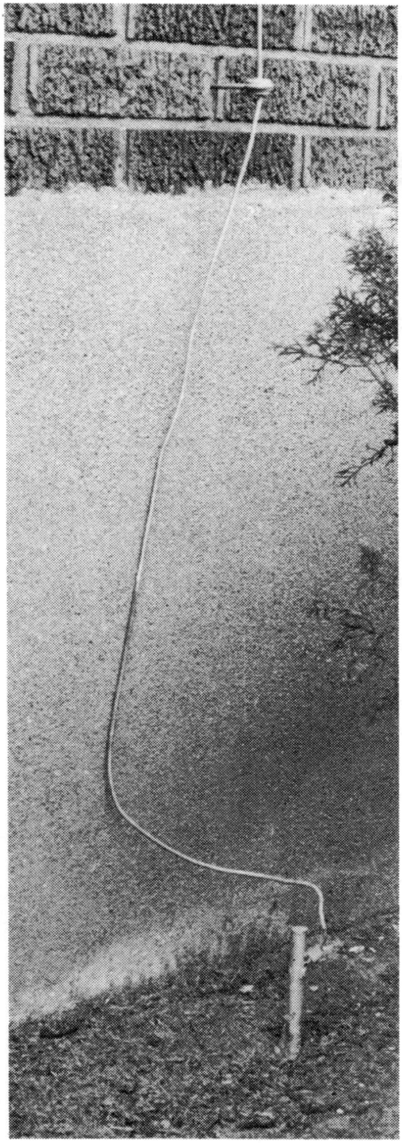

Drive a copper grounding rod into the earth to a depth of 6 or 8 feet. Using about No. 14 AWG wire size, run a lead from the "ground" or "G" terminal of your receiver to the ground rod. Make the ground-lead run as straight and as direct as possible. Use an entry hole a few inches away from the antenna lead-in hole.

DO NOT connect the antenna ground lead to the chassis of any receiver. Use only the "G" terminal beside the antenna terminals. If you have more than one receiver, make a *grounding bus* of No. 14 AWG wire. Connect each receiver's "G" terminal to it. Then connect the grounding bus to the ground rod, as directly as you can. Solder all ground-bus connections.

The antenna lead-in should come as straight to the receiver as it can. It should not touch metal along the way, even though its wire be insulated. This includes metal windows, which make very poor entry points for an antenna lead (and a downright dangerous entry for the grounding lead).

A flat strip comes with some shortwave antenna kits. It lets you bring the antenna lead-in gently under a windowsash. Fahnestock clips (at each end) make poor contact after a short while, so solder the wires to them if you use this strip.

Better yet, drill two entry holes in the house frame. You can buy plastic tubing to line the holes. But that's not really necessary if you drill the holes upward from the outside. Leave a *drip loop* of lead-in wire, supported by a standoff, below the antenna-lead entry.

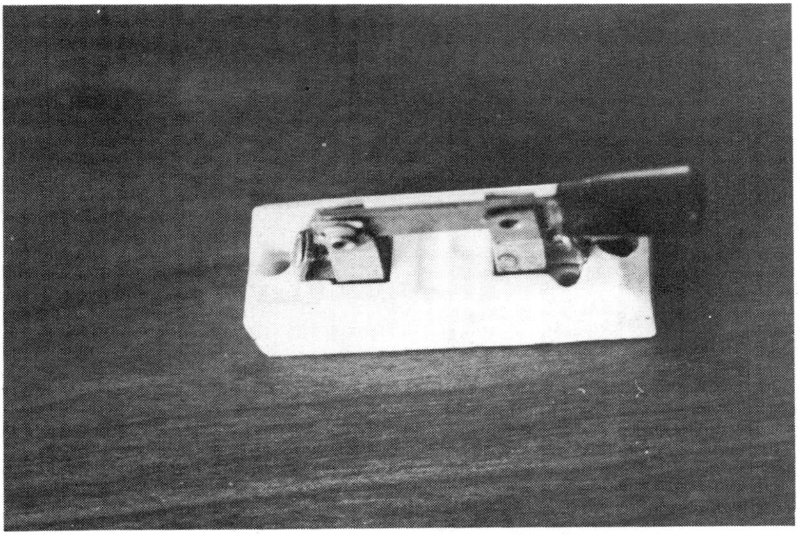

This little device takes the place of a communications lightning arrestor, which may be difficult to find. Very close to the entry hole, but inside the house, connect the hinged end of this knife switch to the antenna ground lead. Make the connection *right at* the ground lead, not through a wire run *to* it. Connect the other end of the knife switch to the antenna lead-in, again right at the wire.

If you're using a doublet antenna and two lead-in wires, buy a two-pole knife switch, and connect *both* hinged ends to the ground lead. Then attach the other two switch terminals to the two lead-in wires.

When your station is not in use, close the knife switch. That grounds your antenna system and prevents it from collecting static lightning charges that might burn the input coils of your receiver. Open the switch for normal operation.

Indoor antennas are one way to get by. Portables have their own antennas built in, collapsible for easy transportability. The receiver at the top has a dipole antenna; the one at the bottom, a monopole. There's no big advantage to one over the other, providing each has been matched correctly to its receiver (by the design engineer).

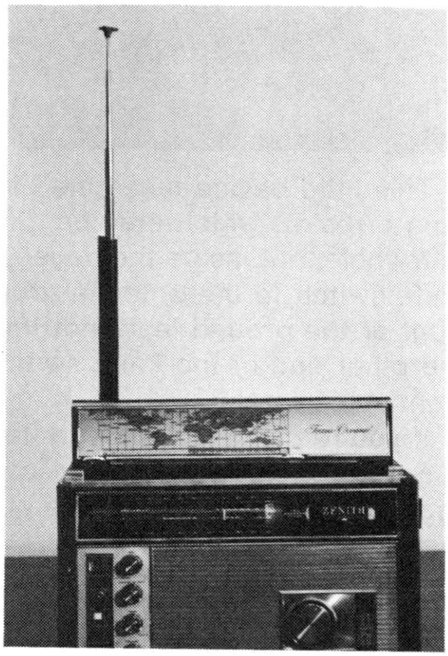

However, for listening to shortwave, the longer the exposed metal antenna rod, the better the reception. The dipoles in the top radio are not really very long; a jack on the side of the receiver lets you plug in a long piece of wire for vastly improved shortwave performance.

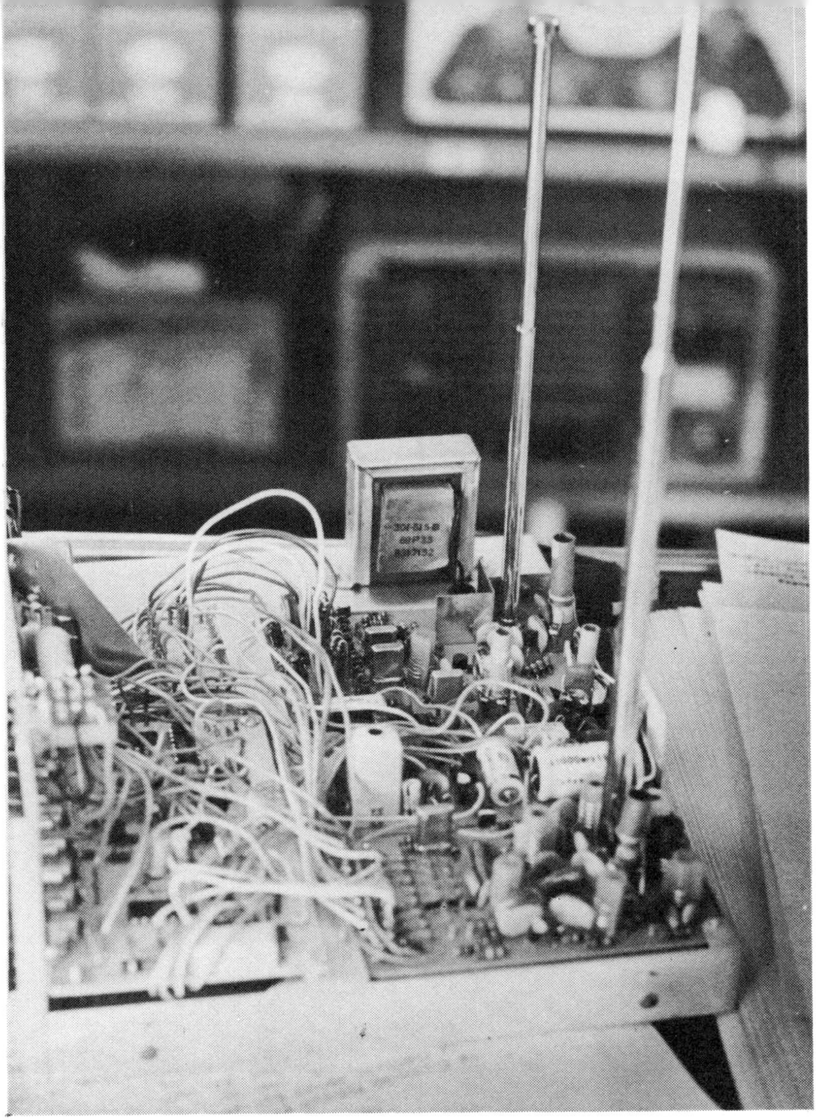

Here's an inside look at one three-band vhf/uhf scanning monitor. You can see the antennas that protrude through the top (see photo, page 71). They're collapsible in the same manner as in portable radios. You'll find that you can improve performance for individual stations by adjusting the lengths of the antennas—longer for frequencies near the low end of the dial, shorter for those near the high end.

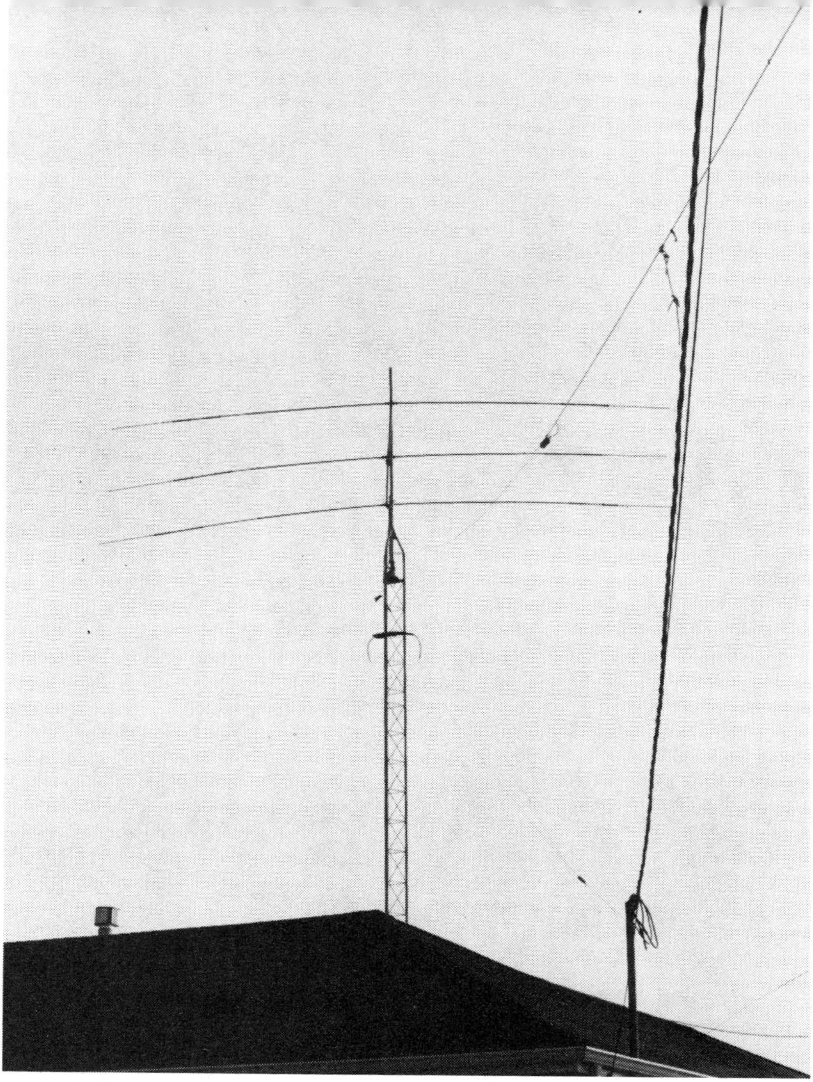

Where you have more than one antenna and more than one receiver, you may want to juggle antennas around a bit. This takes a lot of screwdriver work. Better to install special switching, but that's beyond the nontechnical SWL.

Talk over your needs and wishes with an expert at the communications specialty store nearest you. Or, find a friendly ham operator. Either one can help you work out a switching arrangement for your antennas/receivers. That'll save you unfastening and refastening lead-in wires so much. (You never switch the ground but leave all receivers connected to it at all times.)

Chapter 7

Twirling the Right Knob

Do you know what all the labels below mean? Do you know what to do with the knobs they identify? If not, turn the page. This chapter shows you how to operate a shortwave receiver.

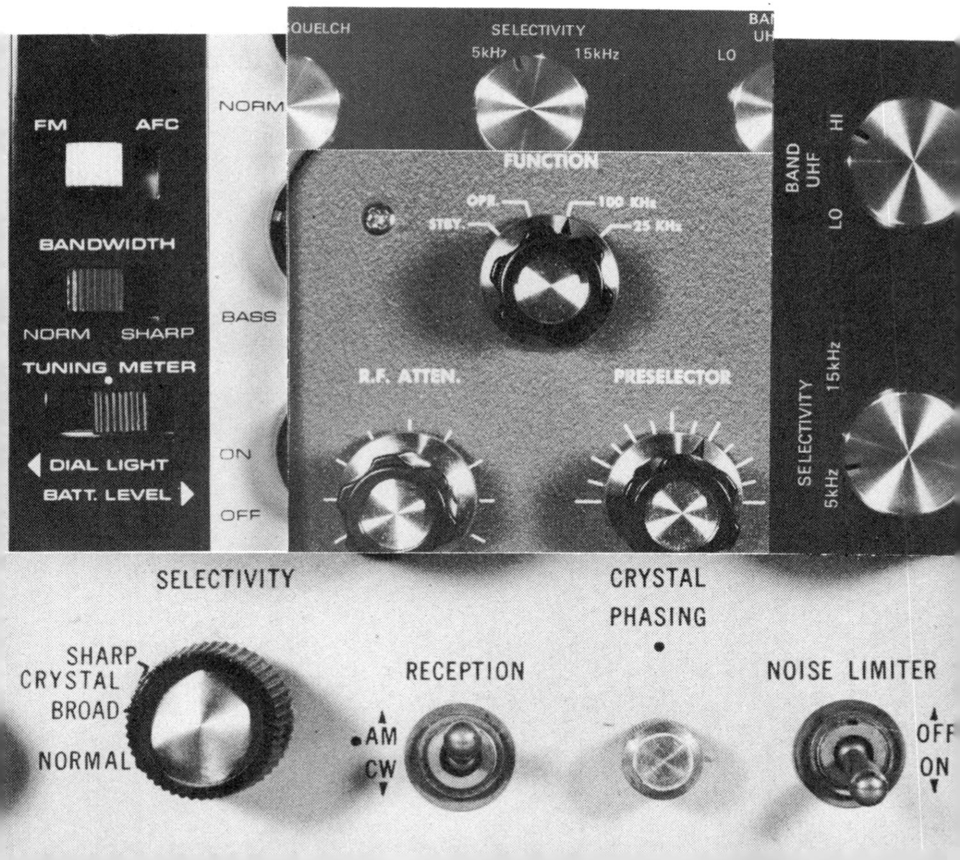

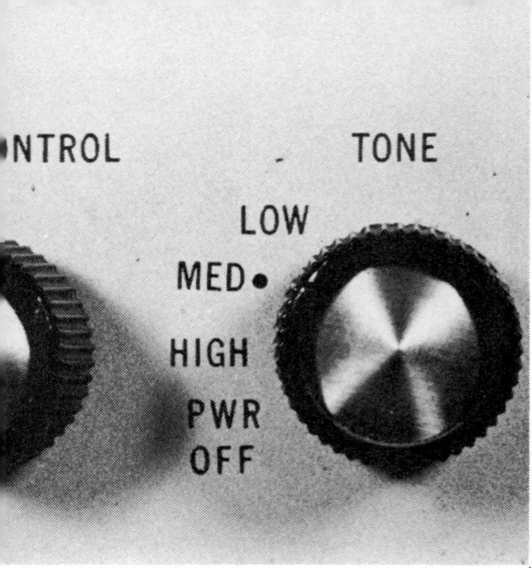

First step involves the power switch. Ordinarily, it accompanies some other control. It might be part of a tone-control switch, with an af (audio frequency) gain control, or actuated by the rf (radio frequency) gain knob. The only clue is the *Power Off* marking.

The power switch applies electricity from the power cord to the circuits. The receiver should start to operate. If it doesn't, and you're sure the plug is in the wall socket and the speaker is hooked up, read the next page.

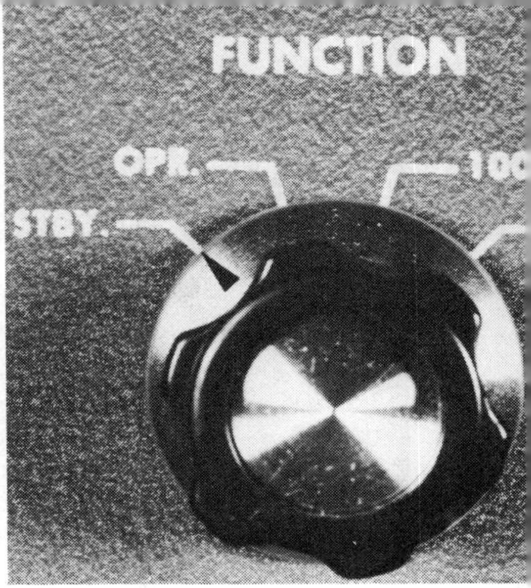

Somewhere on most shortwave receivers, you'll find a switch with the marking *Standby* or some abbreviation of that. The switch must be in the *Operate* position for the receiver to work. On standby, the radio is on but certain circuits are disabled. The receiver then can't pick up and regenerate voice signals from a transmitter operated beside it—as when a ham uses the receiver as part of his station.

A *Mode* switch indicates what kind of reception you expect. It depends on the transmitter at the station you're receiving. The terms a-m, cw, and ssb were explained on pages 34 and 35. The AM position is the normal position to receive broadcast and other a-m stations. The CW position of the Mode switch turns on a bfo so you can receive code. The USB position sets up the receiver to receive the upper sideband, if that's what a single-sideband station is transmitting. LSB selects the lower sideband. Just flip the knob to whichever position gives you clear voice reception.

The *AF Gain* knob on your shortwave set is equivalent to the volume control of an ordinary radio. It turns the sound level up or down to suit your hearing preference.

Most shortwave receivers have another gain control, labeled *RF Gain.* On casual twirling, it might seem to work almost like the af gain control. The difference is this: the rf control affects amplification by the rf and i-f (intermediate frequency) stages in the shortwave receiver. You turn it down slightly if a strong station sounds distorted—a sign these stages are overloaded. For weak stations, you keep the RF Gain knob turned full up.

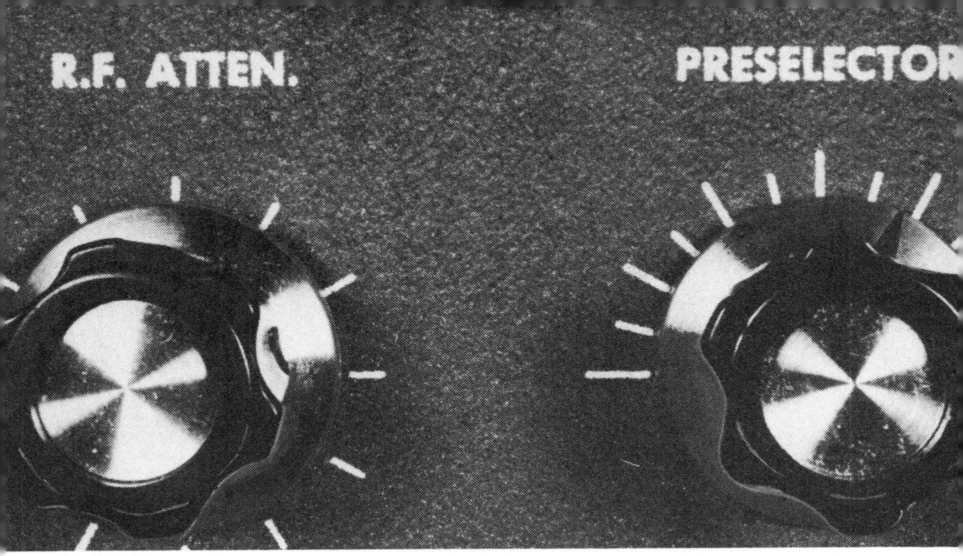

Only a more expensive class of shortwave receiver has either of these controls. The *RF Attenuator* lets you adjust how much signal coming from the antenna reaches the first rf amplifier stage in the receiver. Keep this knob as low as is feasible for the station you're receiving. This protects the first rf amplifier from overload and maybe damage. It's delicate in some receivers.

The *Preselector* allows you to tune that first rf stage for best performance at each particular frequency. You readjust it whenever you change stations. It's a sort of front-end tuning device.

At lower right you see a pair of knobs faintly similar to those above. These are found on a much less costly receiver. The *Antenna Trimmer* accommodates the input circuits of the receiver to whatever length of antenna wire you're using. *Sensitivity* is one more label for a manual (turned by hand) rf gain control.

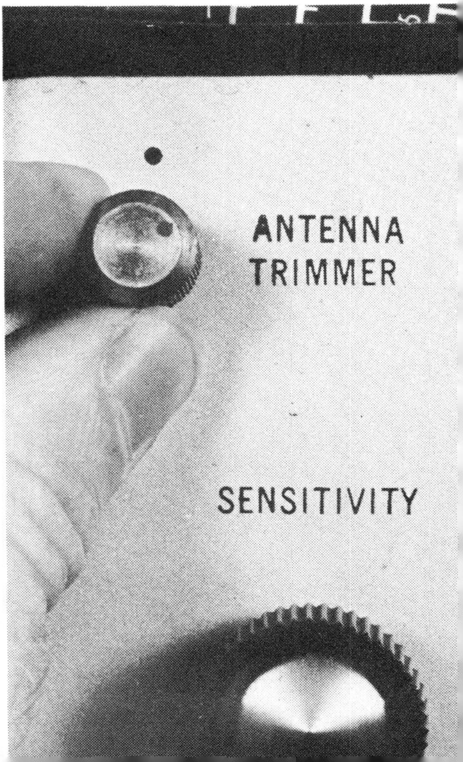

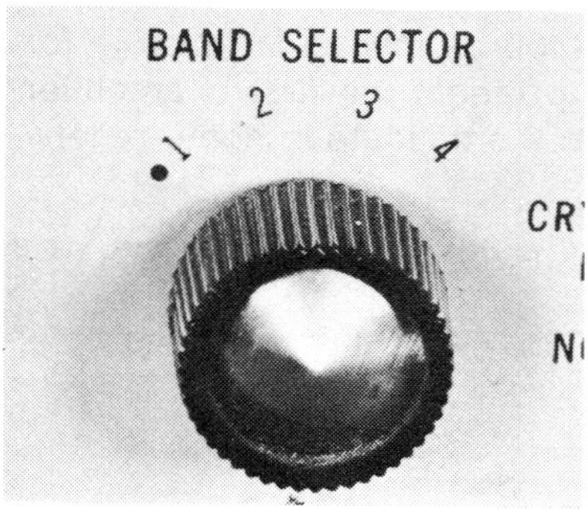

The *Bandswitch* is your key to which line of numbers you look at on the receiver dial. At least, that's true of general-coverage and multiband sets.

The specific-coverage receiver in the bottom photo has a much different tuning system. The numbers by its bandswitching dial list the frequency at the low end of each band covered. For how to use those numbers in dialing a station frequency, see the page opposite.

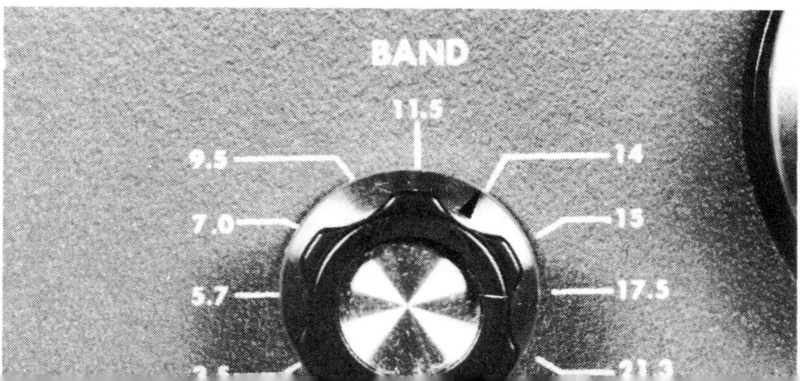

Here's the dial for adding those remaining frequency numerals to the full-megahertz number you select with the bandswitch. This is the main (and only) tuning dial. It adds on frequency (megahertz) digits to the right of the decimal point.

Suppose you want to tune a station at 15.309 MHz. You set the bandswitch dial to 15 (MHz). Then turn the large knob and watch the two dials. The pointer on the straight dial moves slowly past 1, then 2, and finally to 3.

Next watch the crescent-shaped dial while you turn the big knob slowly. Go past the numeral 5, and then four more divisions. The "09" mark comes just before 10. In the photo, you've turned half a division further. The precise frequency (if the dial is calibrated correctly) is 15.3095 MHz. That's about as close to right as you can even expect the station frequency to be; shortwave transmitters in some countries are notoriously imprecise.

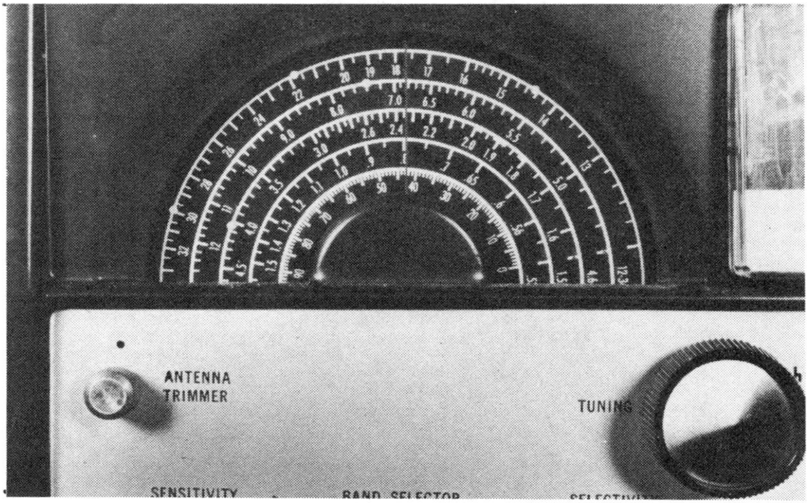

These two pages illustrate another tuning system. It's considerably less accurate, yet you can get pretty near to the station frequency you want and then fish for it.

First you tune the band you want on the main dial. The ham bands are usually marked by small dots. Turn the main tuning dial to that dot on the dial line that corresponds to where you set the bandswitch. This picks approximately the point at which the bandspread dial (facing page) takes over tuning. You can tune stations on the now "calibrated" bandspread with closer accuracy than you could with the main tuning.

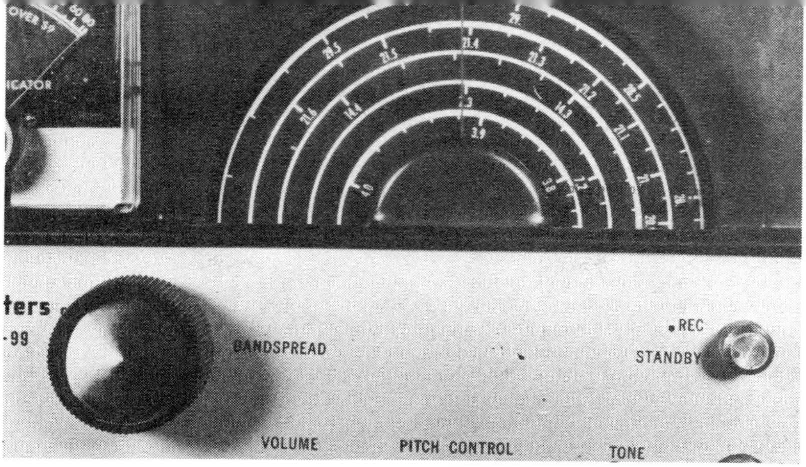

If you want to use the bandspread dial at some frequency outside its marked calibrations, just turn the main dial some even-megahertz (or even-fraction) amount above or below one of the dots—and near the frequency you want to spread out. Then, just add or subtract the difference from the numbers on the bandspread dial.

For example, suppose you want to spread out frequencies between 6.5 and 7.0 MHz. Put the bandswitch at Band 3. The bandspread dial already is marked for about 6.9 to 7.4 MHz, which encompasses the 40-meter ham band. Just turn the *main tuning* dial to 6.8 MHz, which is 0.4 MHz below the dot on that line.

The bandspread dial is shown turned to 7.3. When you subtract the 0.4-MHz error you cranked in on the main dial, the actual receiver frequency is 6.9 MHz. Now you can refine the tuning to the specific station frequency you want, using the Bandspread or Fine Tuning knob. Just remember the 0.4-MHz difference.

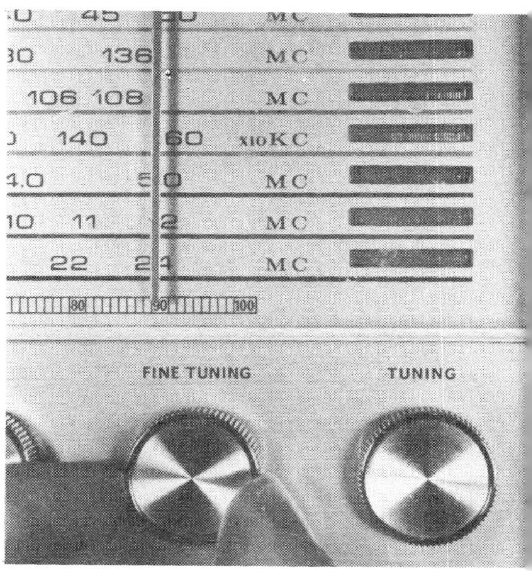

AGC stands for *automatic gain control.* It's a circuit in the receiver that compensates for fluctuations in signal strength. The automatic circuit varies amplification of the rf and i-f stages about like the RF Gain knob does. The agc senses the amount of incoming station signal and automatically prevents overloading. With the agc on, the manual control has no effect.

An elaborate ssb receiver gives you a choice of *Fast* or *Slow* agc. Fast agc works well for a-m reception because the automatic circuit follows the carrier level. For ssb, the automatic circuit might try to follow voice modulation in the sideband since the carrier was eliminated at the transmitter. So . . . the agc must be switched to Slow for ssb reception. For code reception, the agc is best switched off, and rf gain controlled manually.

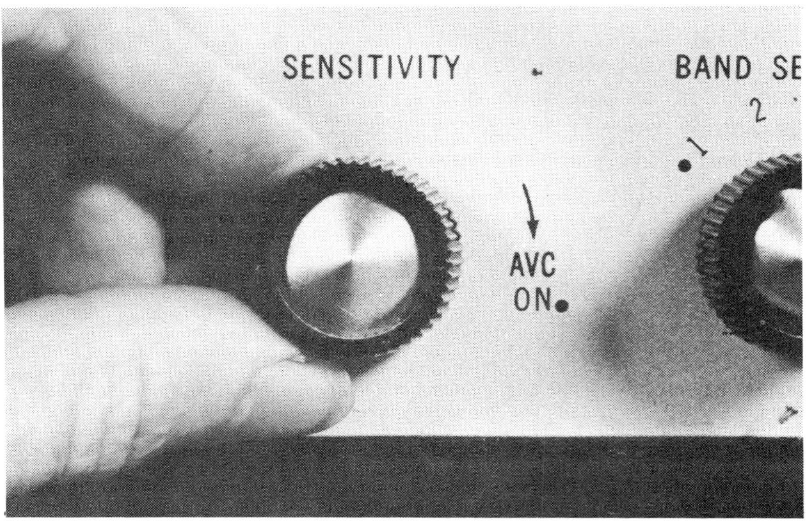

Two knobs on this receiver let you calibrate the dial to pinpoint accuracy. One turns the *Function* switch. The other is the little knob just beneath and to the right of the straight tuning dial. It moves the hairline marker that is over the face of the crescent-shaped dial.

The *Function* knob actuates either a 100-kHz or 25-kHz crystal oscillator inside the receiver. The oscillator's *harmonics* (frequency multiples) fall every 100 kHz (0.1 MHz) or 25 kHz (0.025 MHz) in the band you've chosen with the bandswitch. Turn the main tuning dial close to the frequency you're searching for. You'll hear an audible note from the receiver's speaker.

Carefully turn the main tuning knob until you get a *zero-beat* in the speaker. This means that the audible sound note gets lower and lower in pitch as you turn the dial, until the sound is so low it disappears. If you turn further, the note starts getting higher in pitch again, "on the other side" of the zero-beat point. Set the knob for an exact zero-beat.

Now turn the little knob so the hairline marker is precisely over the "0" reading on the crescent-shaped dial. Or, if the function switch is set to activate the 25-kHz oscillator, the hairline marker can be set to one of that crystal's harmonics.

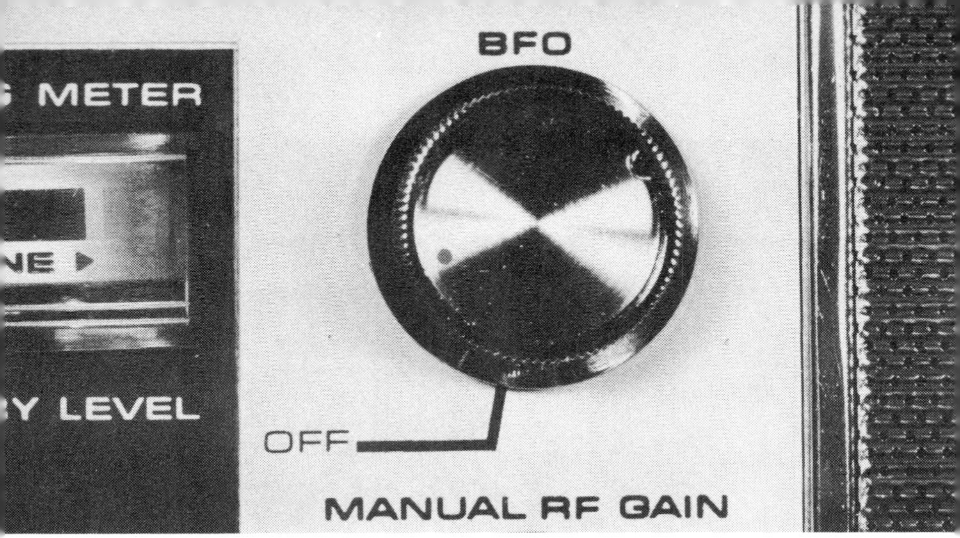

This *BFO* marking, you already know, stands for *beat-frequency oscillator.* On medium- and high-priced receivers, the bfo can be adjusted, so you can alter the pitch of the code tone to suit your ear. The human ear, and many speakers and earphones, show greater sensitivity to some frequencies than to others. For very weak code stations, you need every advantage you can conjure up. Also, a variable bfo lets you sometimes "tune" unwanted code stations up out of range of the speaker or headphones.

The *Pitch* knob on some receivers does the same thing as the bfo dial. The label more truly describes what happens as you vary the bfo frequency.

You can alter code tone pitch with the main or bandspread tuning dials. On a cheap receiver, that's all you have to vary code pitch with. Sometimes, even with a top-notch set, tuning in a code station for optimum reception entails juggling both tuning and bfo frequency. Only experimentation can tell you; there's no hard-and-fast rule.

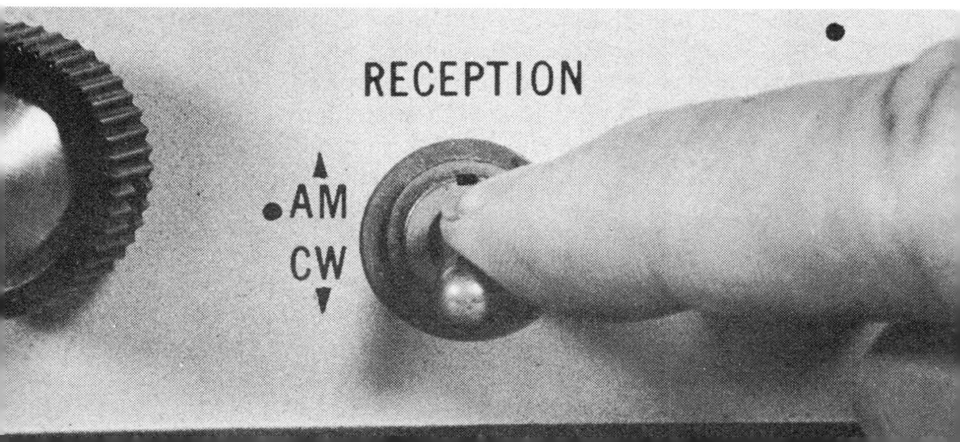

THE INTERNATIONAL CODE

A •–	M ––	Y –•––
B –•••	N –•	Z ––••
C –•–•	O –––	1 •––––
D –••	P •––•	2 ••–––
E •	Q ––•–	3 •••––
F ••–•	R •–•	4 ••••–
G ––•	S •••	5 •••••
H ••••	T –	6 –••••
I ••	U ••–	7 ––•••
J •–––	V •••–	8 –––••
K –•–	W •––	9 ––––•
L •–••	X –••–	0 –––––
Question Mark ••––••		Period •–•–•–
Error ••••••••		Comma ––••––
Wait •–•••		End of Message •–•–•

Learning Morse code can be fun, but it's also work. Here are the dots and dashes, the longs and shorts, the dits and dahs that make up each letter. You can learn by copying code stations. You can buy code records or tapes that offer practice in receiving code groups. You can enroll in a code course that comes with a machine you can speed up and slow down to fit your learning rate. However you go about learning, your shortwave listening will be enriched when you can understand code transmissions as well as listen to a-m or ssb stations.

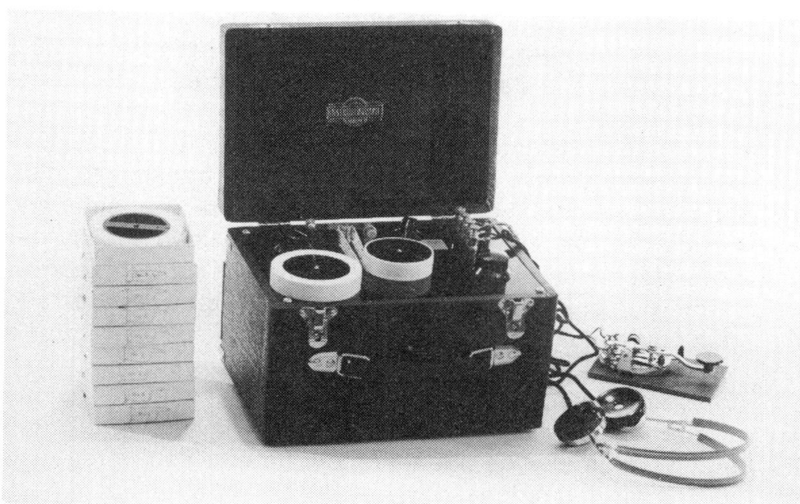

If you can afford a single-sideband receiver, skip these two pages. They explain how to clear up the sound from ssb stations with any receiver that has a broadly adjustable bfo. Well . . . maybe not exactly clear the sound up, but you can decode the "Donald-Duck" chatter enough to make out what's being said. You need patience, a steady hand, and concentration. But there is a way.

Single-sideband signals, in case you've forgotten, consist of only one sideband from what's created when a transmitter modulates a carrier signal. For ssb transmission, the other sideband and the carrier are eliminated. (A-m signals keep the carrier and both sidebands.) A receiver for ssb manages to recover voice from one sideband by supplying a substitute carrier and then demodulating. You can make any receiver do that, after a fashion, using the bfo signal as a substitute carrier.

Start by fine-tuning the ssb station until it's as loud as you can get it with the agc disabled. Then turn the rf gain down low and the af gain (volume) wide open. Flip on the bfo. You're ready to start fishing for that sideband.

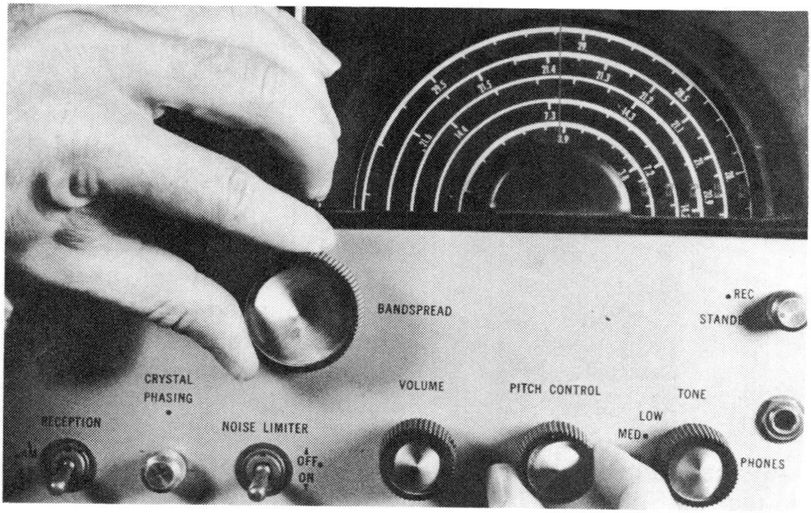

First, merely try tuning the bfo frequency knob (Pitch, in the illustration). The hollow-sounding pitch of the voice rises and falls as you turn the knob back and forth. If the station is tuned about right, at one *critical* setting of the bfo knob you'll be able to make out what the voice is saying.

You may have to move the bandspread dial very slightly. Then try again with the bfo pitch. At some combination of the two, the voice becomes understandable. Strained-sounding, maybe, but intelligible nevertheless.

You can be scientific about your knob-twiddling. Hams, who use ssb more than anyone except ship operators, generally use the lower sideband in the 40- and 80-meter bands and the upper sideband in the 20- and 15-meter bands. So, for the lower sideband, move the bandspread dial almost imperceptibly up-frequency from the station's loudest point. Then turn the bfo pitch back and forth for best clarity. You'll have to do it with care; the point is critical. For an upper sideband, drop the bandspread dial slightly down-frequency from what seems the strongest reception point; try adjusting the bfo pitch again.

When you've struck the right combination of tuning dial setting and bfo pitch, you can understand the ssb operator.

The *selectivity* of a radio was mentioned on page 61. This quality relates to how well a receiver can pick out one station while rejecting stations near it in frequency.

Another term for this is *bandwidth.* This word derives from the fact that a station carrier and its two sidebands occupy a certain tiny amount (or "band") of frequency spectrum. The narrower a receiver's bandwidth, the better it cuts out adjacent stations. The receiver is *selective,* or tunes sharply. Hence the designations "sharp" and "broad" for bandwidth.

A *normal* bandwidth accepts broadcast-type stations. Music, for example, makes sidebands much broader than does voice. *Sharp* bandwidth limits receiver response. The outer limits of music sidebands are lopped off, but voice sidebands come through okay. Bandwidth can be narrowed even further for a single-sideband signal.

In fm receivers like the uhf/vhf monitor, selectivity has a vaguely different function. Some two-way radio transmitters use wideband fm, modulating 15 kHz up and down from center frequency. Others are narrow-band, at 5 kHz. The Selectivity knob on a good receiver lets you pick bandwidth and detector response to suit the station you're hearing.

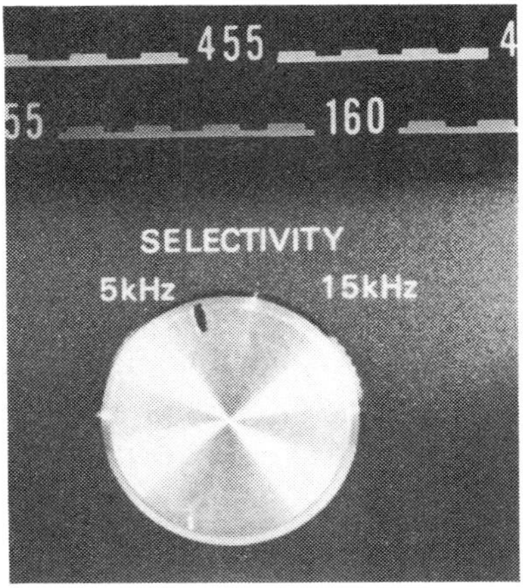

Some shortwave receivers have another way of improving selectivity, particularly for code reception. Bandwidth can be extremely narrow for code, since there's only an interrupted carrier—virtually no sidebands. A *crystal filter* very cleanly blocks out the signal next to whichever one you've tuned in.

To use the crystal filter, you tune the station in, with the bfo turned on. Then flip the filter switch and adjust the *Crystal Phasing* knob to trap out all code tones except the one you're listening to. If need be, readjust bfo pitch slightly. Some filters have a *Broad* and a *Sharp* position. The broad mode lets you trap out a voice station when you find one with a frequency too close to the station you're trying to hear.

Below, you see an older receiver with an *audio boost filter.* You tune in the station and adjust code pitch to a tone you like. Then, you turn up the *Boost* knob about halfway and adjust the *Frequency* knob until the tone from the station you've chosen is the only one amplified. Find a happy medium between Boost and Audio Gain. You will be able to copy the station you want without hearing the ones near it (they have different-pitch tones which are *not* boosted).

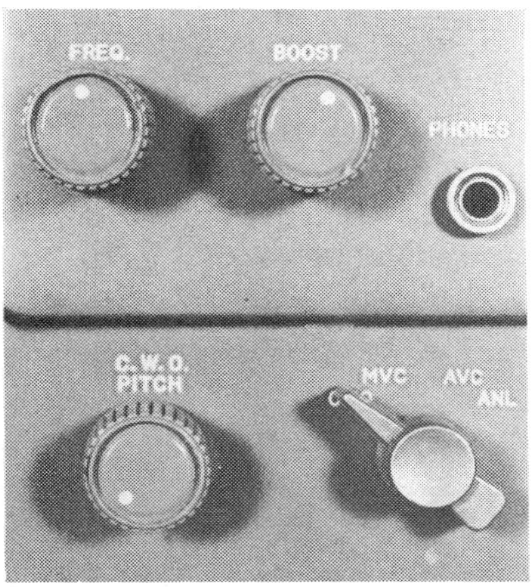

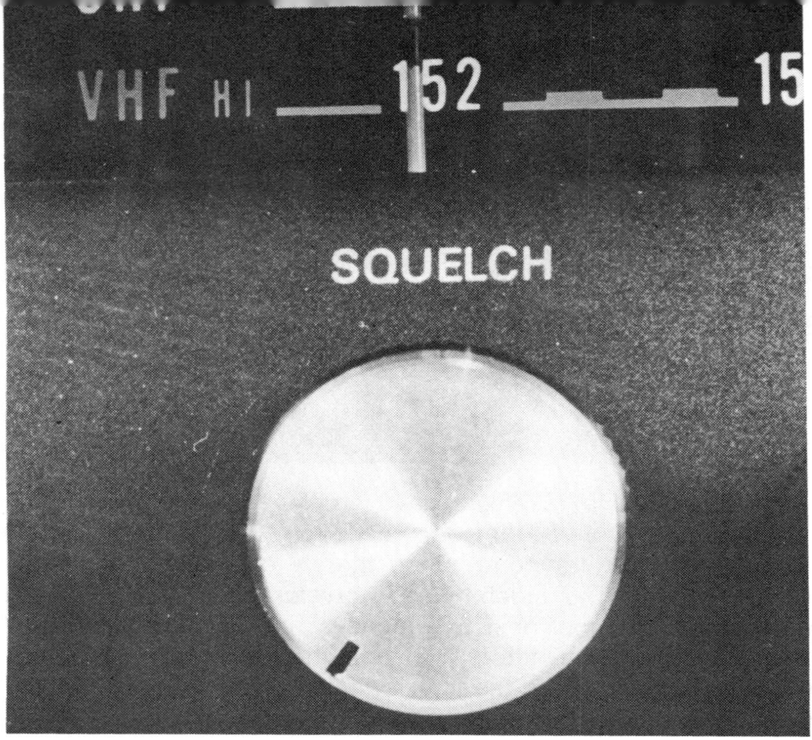

Use of the *Squelch* control on a monitor seems to be a mystery to some listeners. Its job is to eliminate the receiver *background noise* you hear when no station is being received. When the set picks up a transmission, the station signal "quiets" the receiver noise.

Turn the Squelch knob all the way down and the volume control up. Tune away from any station. Now slowly advance the squelch control until the crackling or hissing noise in the speaker stops. Turn the knob just a fraction further for good measure, but not much. If you turn the squelch control too high, you'll miss some signals. From then on, leave the Squelch knob alone. It has been set for optimum sensitivity.

Go ahead and tune the dial for the stations you listen to. If you've tuned right, you will hear the stations when they transmit. Yet the receiver remains silent until they do.

An *automatic noise limiter* (ANL) in some receivers helps shut out certain types of interference noise. It cuts out sharp-sounding impulse noises like automotive ignition sparking. It helps, to some degree, around fluorescent lights.

The noise limiter should be left turned off until you really need it. It decreases the sensitivity of the receiver a little, especially for code reception. You can find additional help with noise by turning the tone control (if the set has one) toward the bass (low) end. Don't go too far with that, though; you'll spoil voice intelligibility.

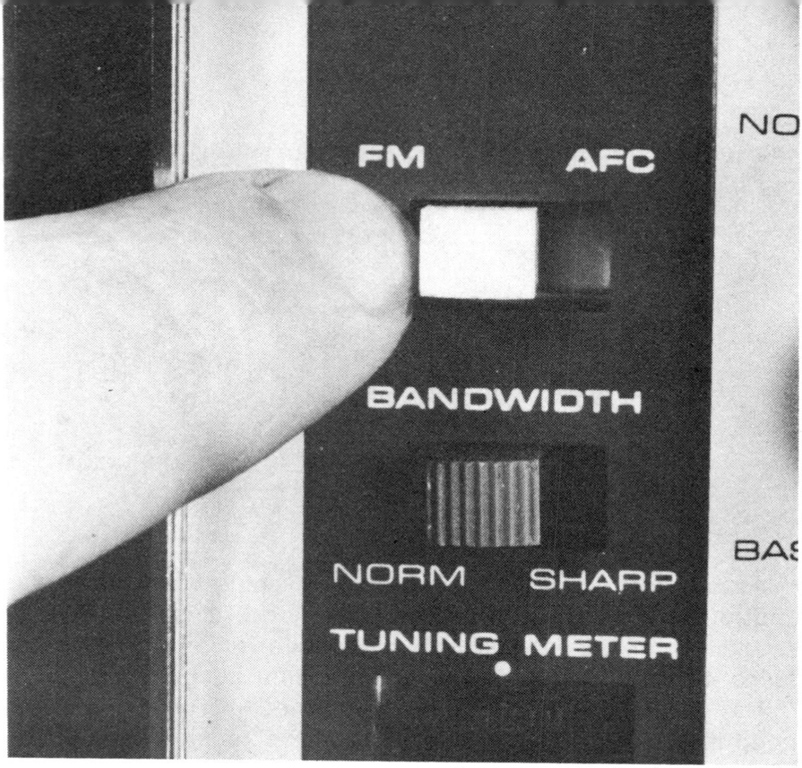

That takes care of just about all the controls and switches you're likely to find on your shortwave receiver. By now you should know how to operate just about any model you choose to buy. Even though the labels may be slightly different and the tuning mechanism not exactly the same, the principles you've seen in this chapter fit them all.

For a windup of this chapter, notice the *AFC* switch on receivers—particularly portables—that tune the fm broadcast band. The term stands for *automatic frequency control.* You tune in a station with the afc turned off, and then activate it. The afc holds the station precisely on-frequency, without drift, for best program reception. Just remember, however, this circuit doesn't operate with any other band on the receiver—only broadcast fm.

Chapter 8

Best Listening Time — Day or Night?

As you dial across the various bands of your shortwave receiver, you'll discover that sometimes you can hardly hear anything from some of them. This natural phenomenon can leave you wondering what's wrong with the receiver. Experience eventually apprises you in a general way that some hours are better for listening than others.

Shortwave propagation (the way radio waves travel) follows a sort of pattern. After lots of guessing and trying, you might figure out the best times for various frequencies. But you don't have to wait. The next few pages explore the dominating influences. From them, you can predict for yourself about what kind of reception to expect in any particular period.

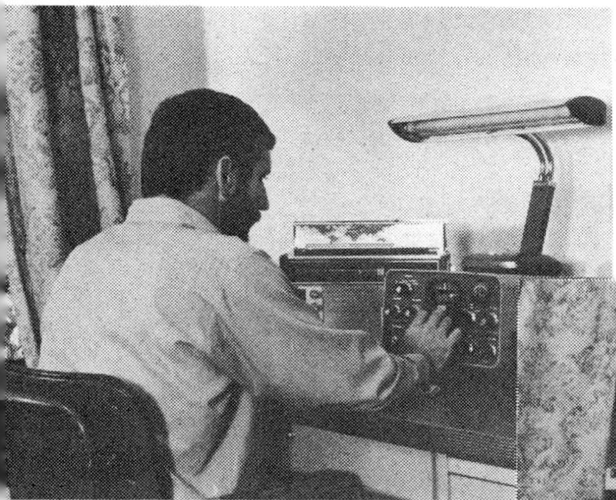

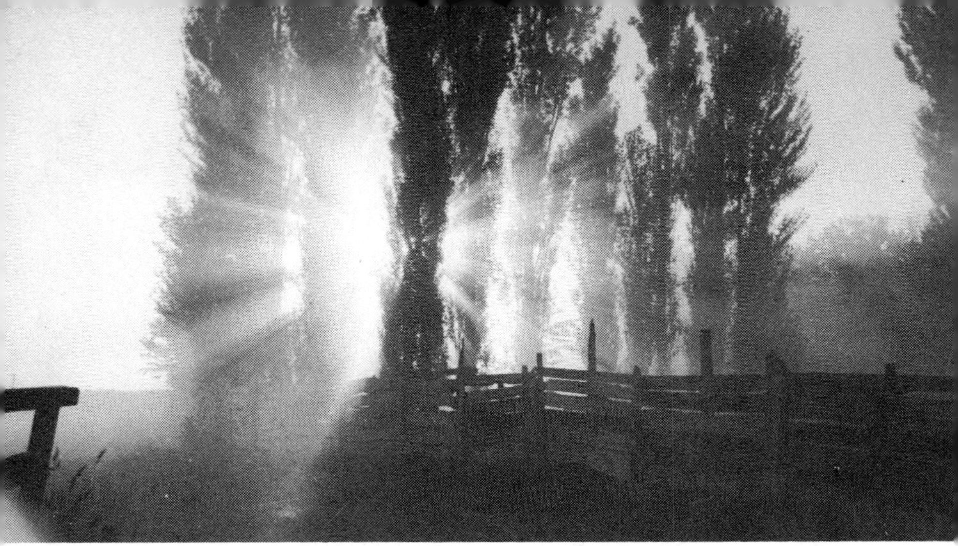

Two key factors in shortwave reception are dawn and sunset. Shortwave listening at these two times of day is unpredictable. The earth's atmospheric layers change radically in regions where sunset and dawn are occurring, upsetting radio propagation.

You should right now orient your thinking away from the usual. Don't use time in terms of the time-of-day where you live. Start considering the world community as a large spinning globe. That lets you recognize instantly that dawn and sunset are both constantly moving. You'll begin seeing the global world as divided into four major time-of-day areas—all continuously moving: (1) before sunset—the afternoon and early evening, (2) after sunset—late evening and nighttime, (3) before dawn—the wee morning hours, and (4) after dawn—the morning and forenoon. For shortwave listening, visualize dawn and sunset as lines of demarcation.

View the earth something like this photo. The dark half is nighttime; the light side, daytime. The earth spins from west to east. As you see it here, sunset is falling over the eastern edge of the United States. In another hour, sunset will lie over the Midwest. An hour after that, in the Rocky Mountains; and in one more hour, on the Pacific Coast.

For the sake of examining the behavior of shortwave signals, imagine that radio signals from 2 through 8 MHz act one way and signals from about 8 through 25 MHz act another.

1. When the transmitting station lies in an area of darkness and the receiving station is in daylight, the higher frequencies are easiest to receive.
2. As the earth turns, and the receiving station moves into the darkness and the transmitter lies in daylight, the lower frequencies travel better.

These are generalities, subject to seasonal variations. But for a beginning, these rules-of-thumb can help you estimate what frequencies you can receive from which parts of the globe, when you (and they) are in areas of light or darkness.

Consider propagation conditions when the earth happens to be in the position illustrated above. It's about midafternoon in the central United States. The sunset line falls about the middle of the Atlantic Ocean, midway between Europe and North America.

Radio transmissions in Europe at frequencies of 9 MHz and above reach the United States with little difficulty. They can be received fairly well throughout most of Canada and the United States mainland. Likewise, receiving stations in western South America can tune in these transmissions. Knowing this, European station operators originating shortwave broadcasts for North and South America time them for post-sunset GMT. That puts those programs over here from midday on.

Just past Greenwich sunset, about 1800 GMT, high-end frequencies are utilized—in the vicinity of 21 MHz. As the sunset line advances toward North and South America, the frequencies chosen are lowered—17, 11, and 9 MHz.

By 2400 GMT and after, Europe experiences dawn. The transmitters emerge into daylight and receiving stations in the Americas move into darkness. From then on, frequencies below 8 MHz seem best: 7, 6, 5, and 4 MHz.

Looking at a similar condition on the other side of the globe, you see Europe in darkness and most of Asia in light. The dawn demarcation lies near the Indian Ocean. The United States has just completed its excursion into darkness. California is experiencing sunset.

The darkened eastern United States receives broadcasts at low-end frequencies best from Asia, which is in daylight. European stations, still in darkness, transmit frequencies around 4 to 8 MHz. The western United States has only sporadic reception from anywhere because sunset is occurring there.

As dawn crosses Europe, their transmitters must move upward in frequency again. Actually, after their dawn, European stations concentrate on domestic broadcasting because shortwave propagation won't improve for several hours.

The seasons have an important effect on radio communications. The general rules for seasonal changes embrace approximately the same frequency segments as those for daylight and darkness. Imagine the 2-8 MHz part of the spectrum as the lower frequencies and about 8 to 22 MHz as the higher frequencies.

Summertime generally favors the high end of the band. The typical day/night considerations apply, but you'll find most reception of lower-end stations disturbed by summer's orientation of the earth with respect to the sun. Summer static (electrical disturbances in the atmosphere) can blot out an otherwise intelligible station. Storms of all kinds are more prevalent in the Northern Hemisphere in summertime, thus limiting reception there. Higher frequencies are more resistant to storm disruption, but they too can be affected during periods of cyclonic activity.

Wintertime offers the best North American reception of any season. Lower frequencies are broadly favored, but good reception of frequencies even well beyond 25 MHz can be expected in the daytime over long distances. The major atmospheric disturbances during this season are the aurora borealis (northern lights) and storms over the northern Atlantic Ocean.

Spring and autumn seem to be "neutral" seasons. Ordinary day/night variations operate consistently in spring and fall, with a diminishing of the disturbing influences that are common to summer and winter.

In all this discussion of radio-wave propagation, two more generalities are important to recognize.

First, north/south communications. You'll find reception from South and Central America best in later afternoon or during the nighttime. Stick to frequencies between 4 and 8 MHz from Central America and from 9 to 25 MHz from South America. This holds true pretty much the year-around, although the winter/summer favoritism of these frequencies does influence.

Second, a "happy medium" falls at about the crossover point in our imagined division of frequencies. For average 24-hour communications, the year-around, frequencies between 7 and 9 MHz are dependable a high percentage of the time.

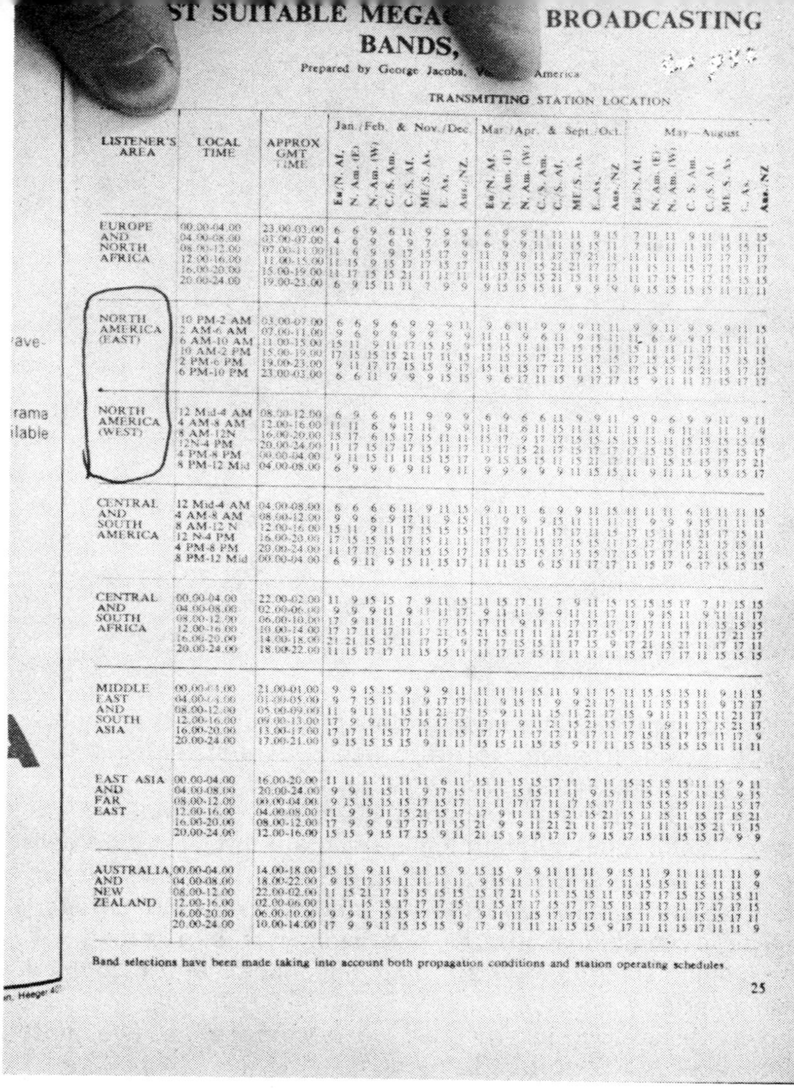

ST SUITABLE MEGA BROADCASTING BANDS,

Prepared by George Jacobs, America

TRANSMITTING STATION LOCATION

Jan./Feb. & Nov./Dec.

LISTENER'S AREA	LOCAL TIME	APPROX GMT TIME	Eu/N. Af.	N. Am. (E)	N. Am. (W)	C./S. Am.	C./S. Af.	ME/S. As.	E. As.	Aus./NZ
EUROPE AND NORTH AFRICA	00.00-04.00	23.00-03.00	6	6	9	6	11	9	9	9
	04.00-08.00	03.00-07.00	4	6	9	6	9	7	9	9
	08.00-12.00	07.00-11.00	11	6	9	9	17	15	17	9
	12.00-16.00	11.00-15.00	11	15	9	15	17	17	15	17
	16.00-20.00	15.00-19.00	11	17	15	15	21	11	11	11
	20.00-24.00	19.00-23.00	6	9	15	11	11	7	9	9
NORTH AMERICA (EAST)	10 PM-2 AM	03.00-07.00	6	6	9	6	9	9	9	11
	2 AM-6 AM	07.00-11.00	9	6	9	9	9	9	9	9
	6 AM-10 AM	11.00-15.00	15	11	9	11	17	15	15	9
	10 AM-2 PM	15.00-19.00	17	15	15	15	21	17	11	15
	2 PM-6 PM	19.00-23.00	9	11	17	17	15	15	9	17
	6 PM-10 PM	23.00-03.00	6	6	11	9	9	9	15	15
NORTH AMERICA (WEST)	12 Mid-4 AM	08.00-12.00	6	9	6	6	11	9	9	9
	4 AM-8 AM	12.00-16.00	11	11	6	9	11	11	9	9
	8 AM-12N	16.00-20.00	15	17	6	15	17	15	11	11
	12N-4 PM	20.00-24.00	11	17	15	17	17	15	11	17
	4 PM-8 PM	00.00-04.00	9	11	15	11	11	15	15	17
	8 PM-12 Mid	04.00-08.00	6	9	9	6	9	11	9	11
CENTRAL AND SOUTH AMERICA	12 Mid-4 AM	04.00-08.00	6	6	6	6	11	9	11	15
	4 AM-8 AM	08.00-12.00	9	9	6	9	17	11	9	15
	8 AM-12 N	12.00-16.00	15	11	9	11	17	15	15	15
	12 N-4 PM	16.00-20.00	17	15	15	15	17	15	11	11
	4 PM-8 PM	20.00-24.00	11	17	17	15	17	15	15	17
	8 PM-12 Mid	00.00-04.00	6	9	11	9	15	11	15	17
CENTRAL AND SOUTH AFRICA	00.00-04.00	22.00-02.00	11	9	15	15	7	9	11	15
	04.00-08.00	02.00-06.00	9	9	9	11	9	11	11	17
	08.00-12.00	06.00-10.00	17	9	11	11	11	11	17	17
	12.00-16.00	10.00-14.00	17	17	11	17	11	17	21	15
	16.00-20.00	14.00-18.00	21	21	15	17	11	17	17	9
	20.00-24.00	18.00-22.00	11	15	17	17	11	15	15	11
MIDDLE EAST AND SOUTH ASIA	00.00-04.00	21.00-01.00	9	9	15	15	9	9	9	11
	04.00-08.00	01.00-05.00	9	7	15	11	11	9	17	17
	08.00-12.00	05.00-09.00	11	9	11	11	15	11	21	17
	12.00-16.00	09.00-13.00	17	9	9	11	17	15	17	15
	16.00-20.00	13.00-17.00	17	17	11	15	17	11	11	15
	20.00-24.00	17.00-21.00	9	15	15	15	15	9	11	11
EAST ASIA AND FAR EAST	00.00-04.00	16.00-20.00	11	11	11	11	11	11	6	11
	04.00-08.00	20.00-24.00	9	9	11	15	11	9	17	15
	08.00-12.00	00.00-04.00	9	15	15	15	15	17	15	17
	12.00-16.00	04.00-08.00	11	9	9	11	15	21	15	17
	16.00-20.00	08.00-12.00	17	9	9	9	17	17	11	15
	20.00-24.00	12.00-16.00	15	15	9	15	17	15	9	11
AUSTRALIA, AND NEW ZEALAND	00.00-04.00	14.00-18.00	15	15	9	11	9	11	15	9
	04.00-08.00	18.00-22.00	9	15	17	15	11	11	11	11
	08.00-12.00	22.00-02.00	11	15	21	17	15	15	15	15
	12.00-16.00	02.00-06.00	11	11	15	15	17	17	17	15
	16.00-20.00	06.00-10.00	9	9	11	15	15	17	17	11
	20.00-24.00	10.00-14.00	17	9	9	11	15	15	15	9

Mar./Apr. & Sept./Oct.

LISTENER'S AREA	LOCAL TIME	APPROX GMT TIME	Eu/N. Af.	N. Am. (E)	N. Am. (W)	C./S. Am.	C./S. Af.	ME/S. As.	E. As.	Aus./NZ
EUROPE AND NORTH AFRICA	00.00-04.00	23.00-03.00	6	9	9	11	11	11	9	15
	04.00-08.00	03.00-07.00	6	9	9	11	11	15	15	11
	08.00-12.00	07.00-11.00	11	9	9	11	17	17	21	11
	12.00-16.00	11.00-15.00	11	15	11	15	21	21	17	17
	16.00-20.00	15.00-19.00	11	17	15	15	21	15	11	15
	20.00-24.00	19.00-23.00	9	15	15	15	11	9	9	9
NORTH AMERICA (EAST)	10 PM-2 AM	03.00-07.00	9	6	11	9	9	9	11	11
	2 AM-6 AM	07.00-11.00	11	11	9	6	11	9	11	11
	6 AM-10 AM	11.00-15.00	15	15	11	11	17	15	15	11
	10 AM-2 PM	15.00-19.00	17	15	15	17	21	15	17	15
	2 PM-6 PM	19.00-23.00	15	11	15	17	17	11	15	17
	6 PM-10 PM	23.00-03.00	9	6	17	11	15	9	17	17
NORTH AMERICA (WEST)	12 Mid-4 AM	08.00-12.00	6	9	6	6	11	9	9	11
	4 AM-8 AM	12.00-16.00	11	11	6	11	15	11	11	11
	8 AM-12N	16.00-20.00	15	17	9	17	17	15	15	15
	12N-4 PM	20.00-24.00	11	17	15	21	17	15	17	17
	4 PM-8 PM	00.00-04.00	9	15	15	15	11	15	21	17
	8 PM-12 Mid	04.00-08.00	9	9	9	9	9	11	15	15
CENTRAL AND SOUTH AMERICA	12 Mid-4 AM	04.00-08.00	9	11	11	6	9	9	11	15
	4 AM-8 AM	08.00-12.00	11	9	9	9	15	11	11	11
	8 AM-12 N	12.00-16.00	17	17	11	11	17	17	11	15
	12 N-4 PM	16.00-20.00	17	17	17	15	17	15	15	11
	4 PM-8 PM	20.00-24.00	15	15	17	15	17	15	15	17
	8 PM-12 Mid	00.00-04.00	11	11	15	6	15	11	17	17
CENTRAL AND SOUTH AFRICA	00.00-04.00	22.00-02.00	11	15	17	11	7	9	11	15
	04.00-08.00	02.00-06.00	9	11	11	9	9	11	11	17
	08.00-12.00	06.00-10.00	17	11	9	11	11	17	17	17
	12.00-16.00	10.00-14.00	21	15	11	11	11	21	17	15
	16.00-20.00	14.00-18.00	17	17	15	15	11	17	15	9
	20.00-24.00	18.00-22.00	11	17	17	15	11	11	11	11
MIDDLE EAST AND SOUTH ASIA	00.00-04.00	21.00-01.00	11	11	11	15	11	9	11	15
	04.00-08.00	01.00-05.00	11	9	15	11	9	9	21	17
	08.00-12.00	05.00-09.00	15	9	11	11	15	11	21	17
	12.00-16.00	09.00-13.00	17	11	9	11	21	15	21	15
	16.00-20.00	13.00-17.00	17	17	11	17	17	11	17	11
	20.00-24.00	17.00-21.00	15	15	11	15	15	9	11	11
EAST ASIA AND FAR EAST	00.00-04.00	16.00-20.00	15	11	15	15	17	11	7	11
	04.00-08.00	20.00-24.00	11	11	15	15	11	11	9	15
	08.00-12.00	00.00-04.00	11	11	17	17	11	17	15	17
	12.00-16.00	04.00-08.00	17	9	11	11	15	21	15	21
	16.00-20.00	08.00-12.00	21	9	9	11	21	21	11	17
	20.00-24.00	12.00-16.00	21	15	9	15	17	17	9	15
AUSTRALIA, AND NEW ZEALAND	00.00-04.00	14.00-18.00	15	15	9	9	11	11	11	9
	04.00-08.00	18.00-22.00	9	15	11	11	11	11	11	9
	08.00-12.00	22.00-02.00	15	17	21	15	11	15	15	11
	12.00-16.00	02.00-06.00	11	15	17	17	15	17	17	15
	16.00-20.00	06.00-10.00	9	11	11	15	17	17	17	11
	20.00-24.00	10.00-14.00	17	9	11	11	11	15	15	9

May—August

LISTENER'S AREA	LOCAL TIME	APPROX GMT TIME	Eu/N. Af.	N. Am. (E)	N. Am. (W)	C./S. Am.	C./S. Af.	ME/S. As.	E. As.	Aus./NZ
EUROPE AND NORTH AFRICA	00.00-04.00	23.00-03.00	7	11	11	9	11	11	11	15
	04.00-08.00	03.00-07.00	7	11	11	11	11	15	15	11
	08.00-12.00	07.00-11.00	11	11	11	11	11	17	17	17
	12.00-16.00	11.00-15.00	11	15	11	15	17	17	17	17
	16.00-20.00	15.00-19.00	11	17	19	17	17	15	15	15
	20.00-24.00	19.00-23.00	9	15	15	15	15	11	11	11
NORTH AMERICA (EAST)	10 PM-2 AM	03.00-07.00	9	9	11	9	9	9	11	15
	2 AM-6 AM	07.00-11.00	11	6	9	9	11	11	11	11
	6 AM-10 AM	11.00-15.00	15	11	11	11	17	15	11	11
	10 AM-2 PM	15.00-19.00	17	15	15	17	21	17	15	15
	2 PM-6 PM	19.00-23.00	17	15	15	15	21	15	17	17
	6 PM-10 PM	23.00-03.00	15	9	11	11	17	15	17	17
NORTH AMERICA (WEST)	12 Mid-4 AM	08.00-12.00	9	9	6	9	9	11	9	11
	4 AM-8 AM	12.00-16.00	11	11	6	11	11	11	11	9
	8 AM-12N	16.00-20.00	15	15	11	15	15	15	15	15
	12N-4 PM	20.00-24.00	17	15	15	17	17	15	15	17
	4 PM-8 PM	00.00-04.00	11	11	15	15	15	17	17	21
	8 PM-12 Mid	04.00-08.00	11	9	11	11	9	15	15	17
CENTRAL AND SOUTH AMERICA	12 Mid-4 AM	04.00-08.00	11	11	11	6	11	11	11	15
	4 AM-8 AM	08.00-12.00	11	9	9	9	15	11	11	11
	8 AM-12 N	12.00-16.00	17	15	11	11	21	17	15	11
	12 N-4 PM	16.00-20.00	17	17	17	15	21	15	15	11
	4 PM-8 PM	20.00-24.00	15	17	17	11	21	15	15	17
	8 PM-12 Mid	00.00-04.00	11	15	17	6	17	15	15	15
CENTRAL AND SOUTH AFRICA	00.00-04.00	22.00-02.00	15	15	15	17	7	11	15	15
	04.00-08.00	02.00-06.00	11	9	15	11	9	11	11	17
	08.00-12.00	06.00-10.00	17	17	11	11	11	15	15	15
	12.00-16.00	10.00-14.00	17	17	11	17	11	17	21	17
	16.00-20.00	14.00-18.00	17	21	15	21	11	17	17	11
	20.00-24.00	18.00-22.00	15	17	17	17	11	15	15	15
MIDDLE EAST AND SOUTH ASIA	00.00-04.00	21.00-01.00	11	15	15	15	11	9	11	15
	04.00-08.00	01.00-05.00	11	11	15	15	11	9	17	17
	08.00-12.00	05.00-09.00	15	9	11	11	15	11	21	17
	12.00-16.00	09.00-13.00	17	11	9	11	17	15	21	15
	16.00-20.00	13.00-17.00	17	15	11	17	17	11	17	9
	20.00-24.00	17.00-21.00	15	15	15	15	15	11	11	11
EAST ASIA AND FAR EAST	00.00-04.00	16.00-20.00	15	15	15	15	11	15	9	11
	04.00-08.00	20.00-24.00	11	15	15	15	11	15	9	15
	08.00-12.00	00.00-04.00	11	15	15	15	11	11	15	17
	12.00-16.00	04.00-08.00	15	11	15	11	15	17	15	21
	16.00-20.00	08.00-12.00	17	11	11	11	15	21	11	15
	20.00-24.00	12.00-16.00	17	15	11	15	15	17	9	9
AUSTRALIA, AND NEW ZEALAND	00.00-04.00	14.00-18.00	15	11	9	11	11	11	11	9
	04.00-08.00	18.00-22.00	11	15	15	11	15	11	11	9
	08.00-12.00	22.00-02.00	15	17	17	15	15	15	15	11
	12.00-16.00	02.00-06.00	11	15	17	11	17	17	17	15
	16.00-20.00	06.00-10.00	15	11	15	11	15	15	17	11
	20.00-24.00	10.00-14.00	17	11	11	15	17	11	11	9

Band selections have been made taking into account both propagation conditions and station operating schedules.

25

You can obtain propagation forecasts that guide you to a more specific judgement of what to expect during particular months and hours. If you buy the World Radio-TV Handbook (see page 145), one page contains a comprehensive propagation forecast for the entire year.

Local weather conditions alter the dependability of any long-range forecasts. Nevertheless, once you know the principles described in the preceding pages, the forecast chart becomes exceptionally informative.

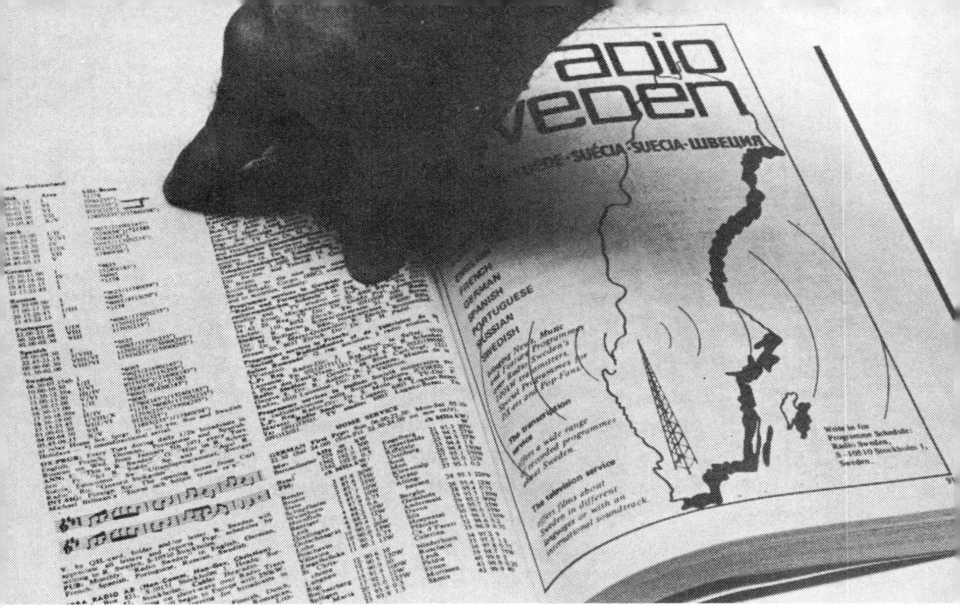

With a forecast chart and a time schedule for the stations you want to listen to, you can plan your shortwave activities. Check the languages too; no point in listening to a station you can't understand (unless for QSL attempts). Most overseas stations broadcast in English at certain times.

Note also in the schedules which direction each broadcast is beamed. It's futile to hunt a Radio Free Europe broadcast that's beamed from West Germany into the Soviet Union. Yet, you might intercept one aimed for Latin America.

Dozens of overseas broadcast stations will gladly put you on their mailing list to receive regular schedules. That'll help you keep up with time and frequency changes as they occur throughout the year. Get addresses from the World Radio-TV Handbook.

Radio waves between 25 and 35 MHz behave differently from those discussed so far. With high power, they bounce around the world. The word "bounce" describes their behavior literally. The ionosphere reflects these signals, angling them back toward the earth. The phenomenon is called *skip.*

Hams use this oddball propagation for worldwide communications in their 10-meter band. CB operators, whose transmitters can legally put out only 4 watts of radio signal, nevertheless "sneak" illicit communications far over the horizon when skip conditions are good. Skip reception can also disrupt CB communications. If skip is too strong, reception of distant stations may block any but very short-distance local communications.

The photo shows a CB operator listening to stations bouncing in from California and from Florida; he's in the north-central part of the country. The law prevents him from answering them. His communications are limited to stations not more than 150 miles away—too near for skip and too distant for his "ground wave" to reach.

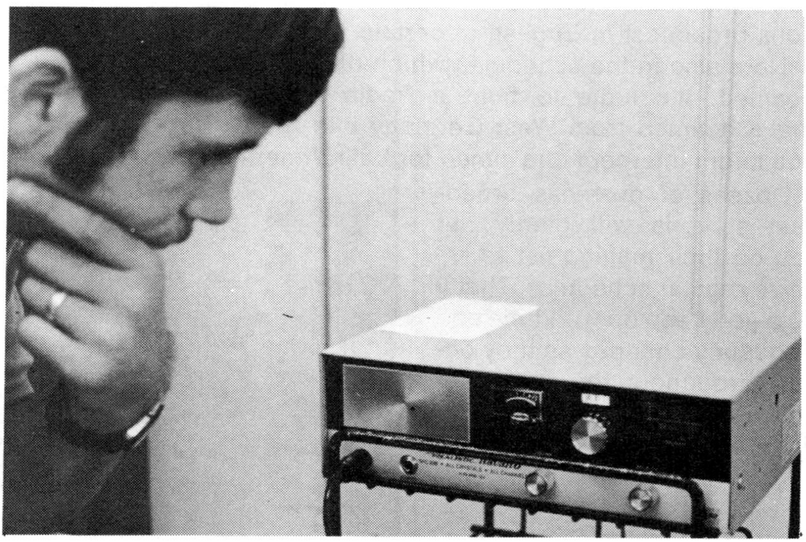

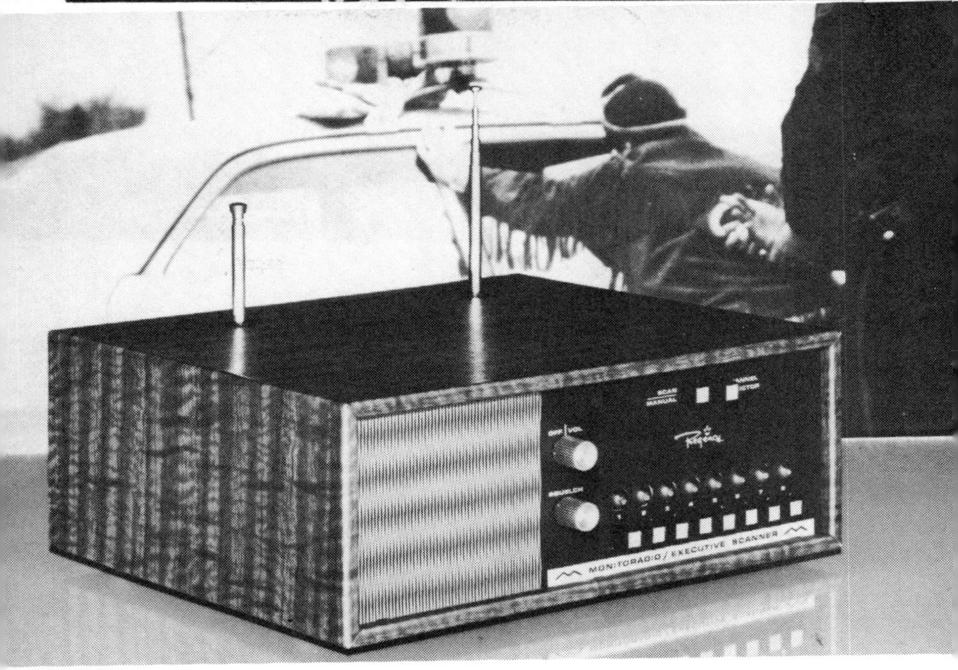

Police, fire, aircraft, and other vhf and uhf communications remain strictly local. During periods of high sunspot activity (explosions on the sun), you may hear stations in the 3- to 50-MHz vhf band bouncing in from great distances—especially from Mexico and South American countries. The sunspots are largely responsible for vhf skip, just as they are for CB skip.

Such reception is a fluke, usually taking place at night. Once in a great while, skip occurs in the 150- to 175-MHz band. At uhf frequencies, never.

Your reception of riverboats and ocean ships depends on what frequency they're using. For communicating over long distances, offshore vessels use frequencies near 12, 18, and 22 MHz in the daytime. Evenings, they more likely switch to 6 or 8 MHz, and then to 4, 6, or 8 MHz at night. The mode of transmission will probably be ssb or cw, sometimes a little of both.

Riverboats keep communication schedules with their home offices sometimes, on 8 or 12 MHz in the daytime and 4 or 6 MHz at night. In coastal waters, either kind of vessel uses 2-3 MHz. At all these frequencies, transmissions usually are ssb. A few older vessels have a-m transmitters, but that ends by 1976.

Near shore, all vessels are required to use vhf-fm. This applies to boats on rivers, whether they're commercial towboats or pleasure craft. The 156- and 160-MHz frequencies are vhf and are good only for short distances. Skip in this band is almost nil. So don't expect to hear these boats unless you live near a waterway, lake port, or ocean port.

Chapter 9

Exploring the Field of Shortwave

Here you see the most complete schedule guide in the world. It's issued annually. You can buy it in December for the following year.

The World Radio-TV Handbook (dubbed "the WRTH") is published in Hvidovre, Denmark. You can buy it in the United States from Gilfer Associates, P.O. Box 239, Park Ridge, NJ 07656, or from Glen Mueller, Billboard Publications Inc., 2160 Patterson Street, Cincinnati, OH 45214. Write for current price; the 1972 issue sold for $6.95.

Books, like this one you're reading, constitute a quick way to learn about SWLing. Any kind of book on communications brings you a clearer understanding of what you're listening to and why you can. Buy books direct or at your favorite electronics store. Here are some of the companies that offer books especially for shortwave listeners:

Howard W. Sams & Co., Inc., Indianapolis, IN 46268

The American Radio Relay League, Newington, CT 06111

Gilfer Associates, Inc., P.O. Box 239, Park Ridge, NJ 07656

SpeeDX, Box 321, Santa Ana, CA 92702

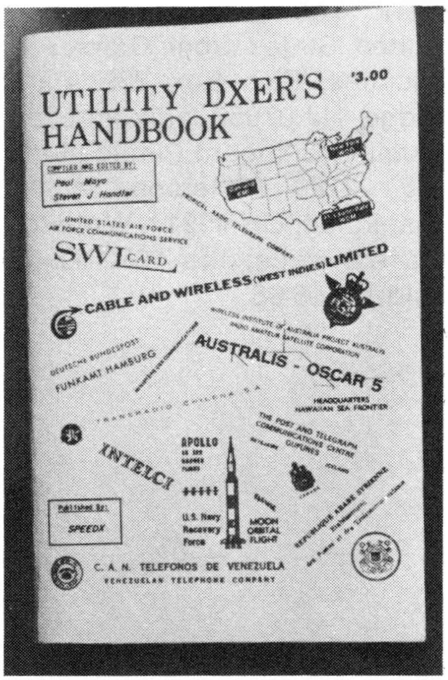

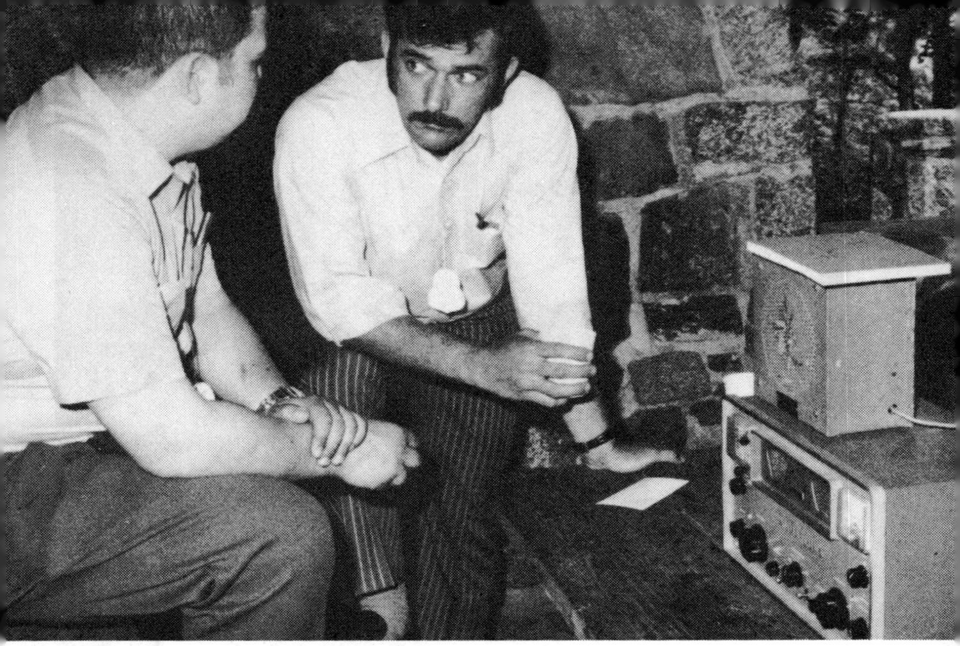

Various clubs publish bulletins that let you know what's being heard, what other SWLs are doing, and what club activities are coming up. Above, two members discuss DXing at an outing of the Newark News Radio Club—one of the country's oldest. Some of the better-known shortwave listener clubs are:

Newark News Radio Club, 215 Market, Newark, NJ 07101
North American SW Assn., Box 8452, Charleston, WV 25303
National Radio Club, 116 Walpole St., Walpole, MA 02081

Clubs come and go. For a list of others currently in operation, drop a postcard to:

Association of North American Radio Clubs, 22 Country Way, North Haven, CT 06473

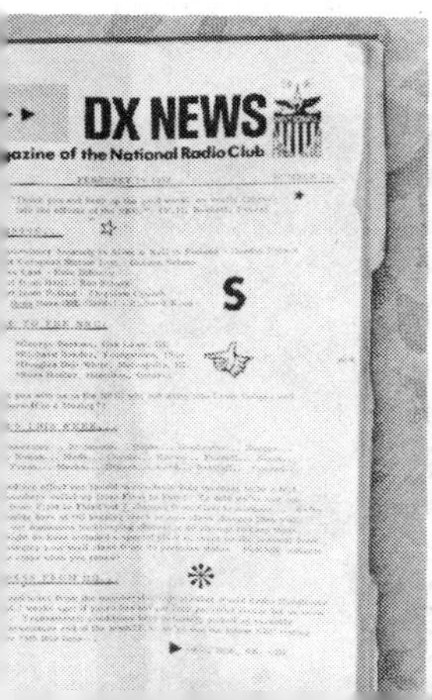

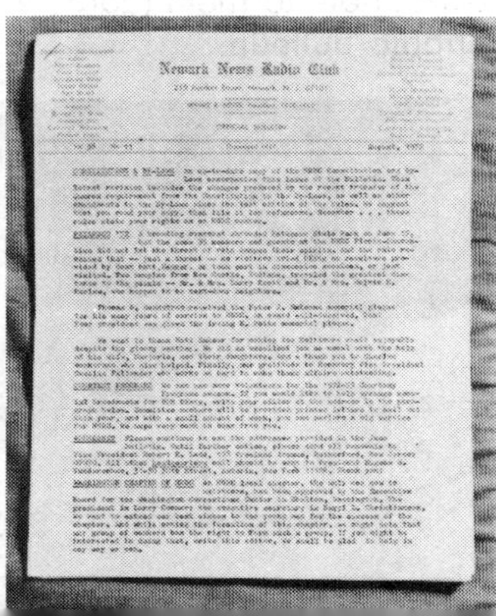

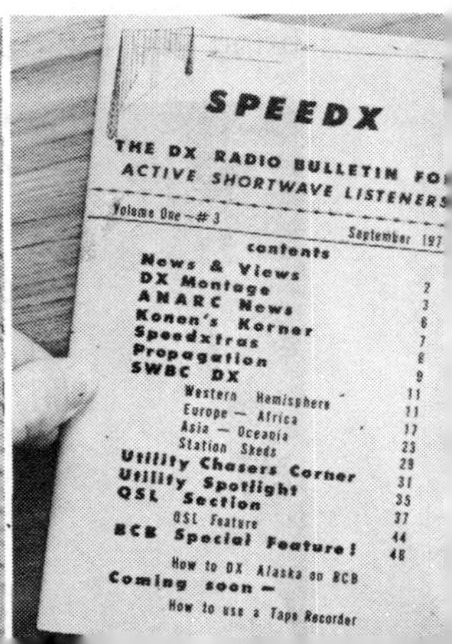

You can form your own local shortwave listener club. One city has such a club built from a nucleus of four enthusiastic shortwave listeners. Each member has different interests, but they swap information to help one another.

Many foreign radio broadcasters have organized clubs of their own. You might join one or more of those. Their club bulletins keep you up-to-date with regard to shortwave doings in that country.

The best-known supplier of shortwave miscellany seems to be Gilfer Associates. A note to them (address on page 145) will bring you their periodic bulletin.

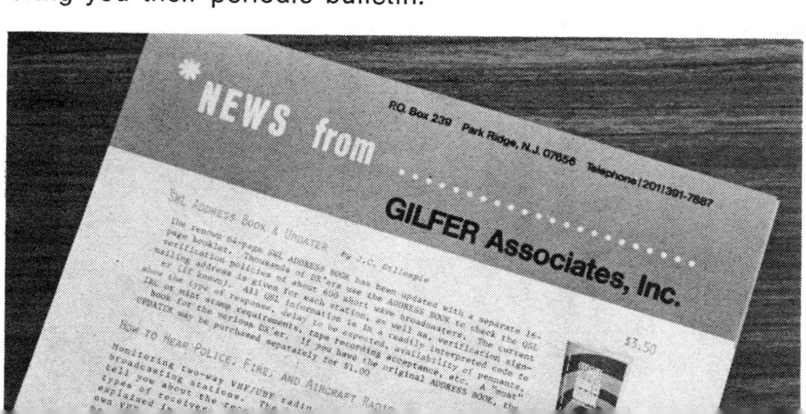

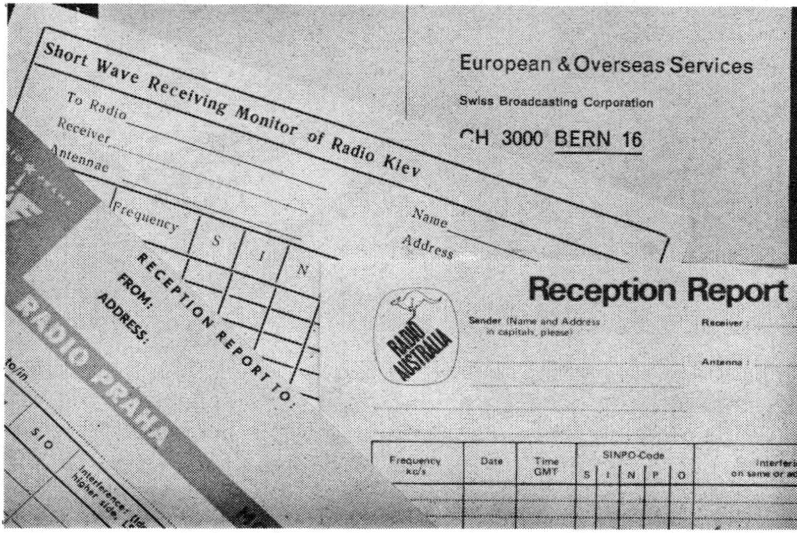

You can become so involved with clubs in other countries that you may be appointed official reception monitor for their major stations. Many of them supply special forms you can use in reporting what kind of signals you receive.

In reporting, you generally use a special numerical rating called the SINPO (sometimes, SINFO) code. The chart reproduced below indicates how the numbers stack up and what each letter in the SINPO acronym stands for. As you can see from some of the report forms, equally important in the report are frequency, date, and GMT.

Signal strength		Interference (Other Stations)		atmosph. Noise (Static)		Propagat. disturb. (Fading)		Overall merit	
5	excellent	5	none	5	none	5	none	5	excellent
4	good	4	slight	4	slight	4	slight	4	good
3	fair	3	moderate	3	moderate	3	moderate	3	fair
2	poor	2	severe	2	severe	2	severe	2	poor
1	barely audible	1	extreme	1	extreme	1	extreme	1	unusuable

You'll want to keep your own log of what you hear and definitely identify on a particular night. You can buy pads of SWL logging sheets already ruled for the information you should log. A carefully kept log helps you build a store of knowledge about reception conditions. Likewise, you can keep track of what stations you have QSL cards from, have asked to send QSLs, or have heard but not requested a QSL from.

The *Shortwave Listeners Guide,* published by Howard W. Sams & Co., Inc., contains log pages at the back, in addition to listing shortwave stations, frequencies, times of broadcast, etc.

But you don't need special forms to keep a log. The one below does for its owner everything necessary for accurate record-keeping. And, of course, no one requires that you keep any log. If your enjoyment is merely in listening and not in accumulating the credit for hearing this or that hard-to-get station, forget a log.

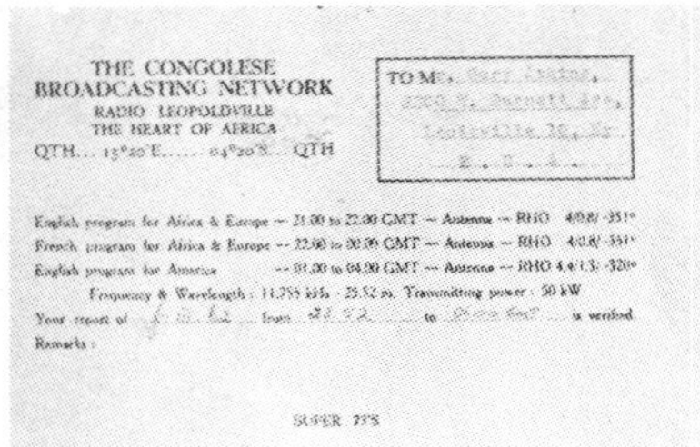

Logging your prize stations is fine. But real proof that you truly did receive them when you said rests in a QSL card from the station in question. Certain stations are difficult to pick up. A verified reception draws prestige in SWL circles. Plus, an unusual QSL boosts your own morale, as it hangs there on the wall of your shack. Some SWLs collect the foreign stamps that bring the QSL cards.

Almost every serious SWL has at least one "special" QSL. Don Williams beams when he describes the day he earned the Radio Mauritania card above. And shortwaver Gary Atkins covets his no-longer-possible QSL from what was once the Belgian Congo. The African and Russian (Moscow) cards belong to an SWL who has since extended his communications knowledge and earned a ham license.

To obtain QSLs, you need only report to a station that you heard it, describe the program you heard, and give the date, GMT, and frequency. Station operators welcome comments and like to see SINPO readings. Be sure to request that they QSL; just type PSE QSL (please acknowledge receipt) at the end of your report.

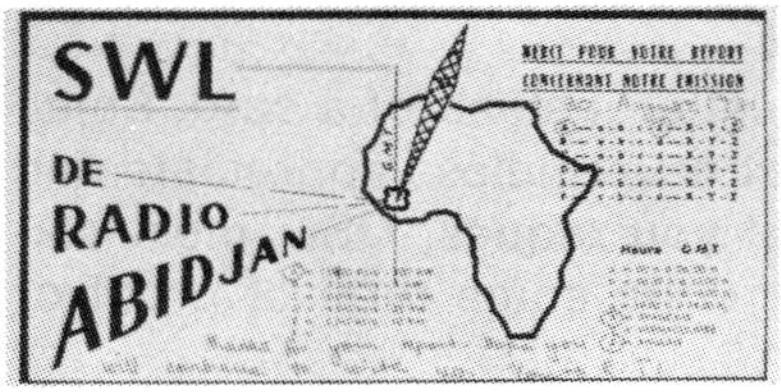

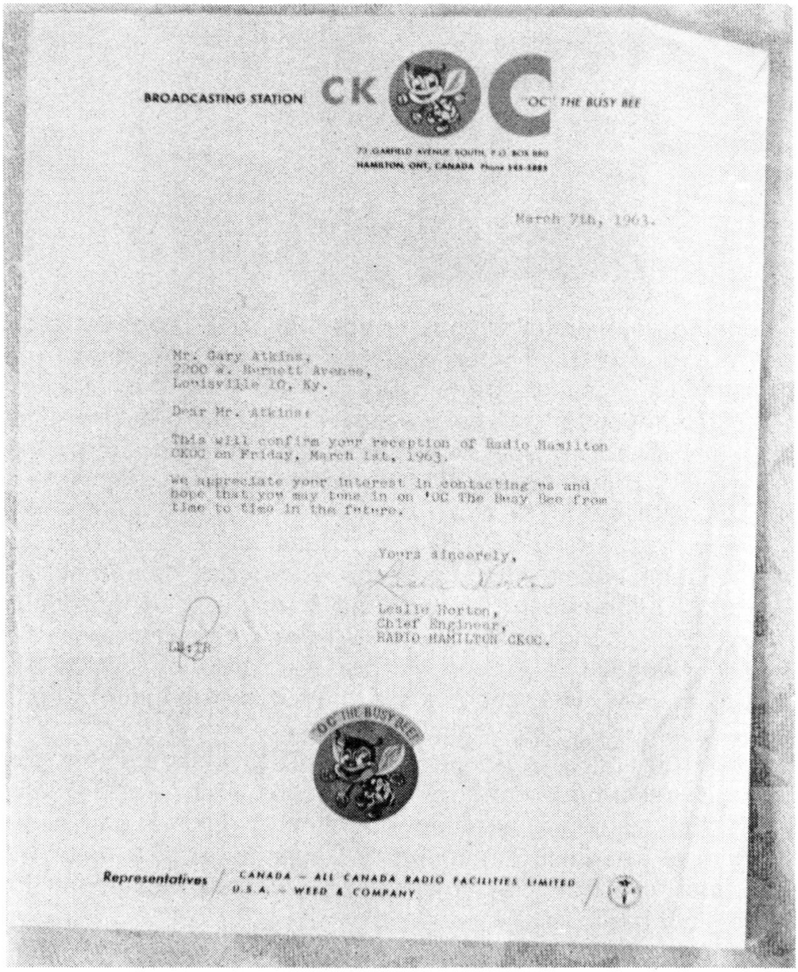

BROADCASTING STATION CK OC "OC" THE BUSY BEE

HAMILTON, ONT., CANADA

March 7th, 1963.

Mr. Gary Atkins,
2200 W. Burnett Avenue,
Louisville 10, Ky.

Dear Mr. Atkins:

This will confirm your reception of Radio Hamilton CKOC on Friday, March 1st, 1963.

We appreciate your interest in contacting us and hope that you may tune in on 'OC The Busy Bee from time to time in the future.

Yours sincerely,

Leslie Horton,
Chief Engineer,
RADIO HAMILTON CKOC.

LH:IR

"OC" THE BUSY BEE

Representatives / CANADA — ALL CANADA RADIO FACILITIES LIMITED
U.S.A. — WEED & COMPANY

You may receive a verification letter instead of a card. Some stations don't have QSL cards, but nevertheless appreciate and acknowledge reception reports. If you request QSL, most station operators make an effort to return some form of proof or verification.

Some stations mail pennants when you consistently monitor and report their transmissions. A few use pennants in lieu of QSL cards.

Jim Lovell has gathered pennants from some of the top shortwave stations of the world. His room-corner station has proved one of the more effective in his part of the country. It's not overly elaborate but does sport refinements that make it special.

Most cherished among his equipment is a frequency calibrator (an Army surplus BC-221 frequency meter). With it, he can pinpoint incoming signals right down to the fraction of a kilohertz, identify an unknown station by its frequency, or tune his receiver with precision ordinarily found only in sets far more expensive.

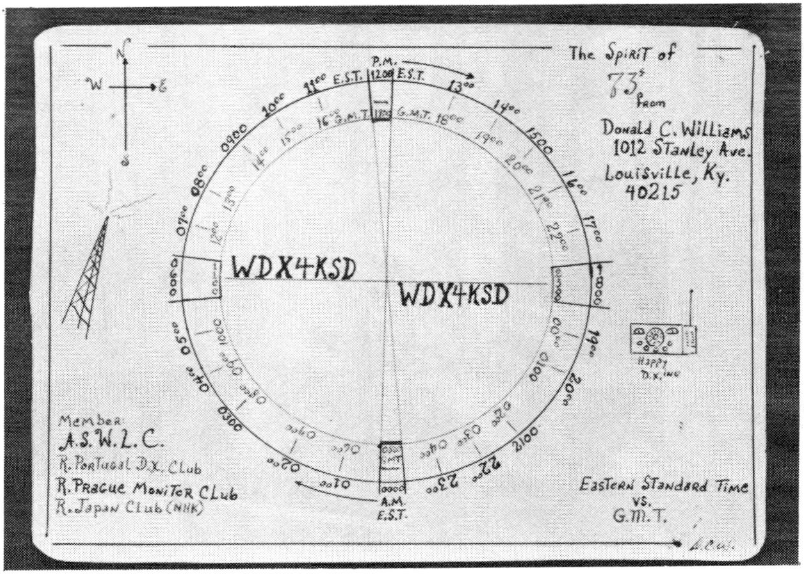

Here's a QSL in reverse. Don Williams, a particularly enthusiastic SWL, draws up his own outsize QSL card in the form of a GMT/EST conversion chart. He calls his monitoring station by the letters WDX4KSD. (W4 is the official call-letter prefix for hams in FCC Region 4.) Don sends these cards to fellow SWLs from time to time. Just one more way of keeping in touch with the magic world of shortwave listening.

Magic world? Sure it is. How else could you sit at an 8×10×16 box in a kitchen corner in Louisville, Kentucky, and know the politics of the world firsthand, learn smatterings of five languages directly from natives of the countries where the languages are spoken, become pen-pals with top broadcasters on the other side of the world (or drive your wife to distraction over hours spent glued in front of "that knob-covered monstrosity")? Shortwave listening is the only way.

Chapter 10

Shortwave Stations to Try

The newcomer to shortwave listening usually does best with English-speaking stations. This is a partial list of countries that broadcast at least some of the time in English. They carry news, propaganda, and entertainment. Schedules and frequencies are shuffled around considerably. Consult the *WRTH* or *Shortwave Listener's Guide* for current listings. Or write directly to stations that interest you; they'll gladly send schedules (addresses in WRTH).

This chapter is divided into two parts. First, some broadcast stations of different countries are listed. Then follows certain radio services in the United States and the bands they use. Some of the stations in the second part can be picked up on vhf/uhf monitors but most require a tunable shortwave receiver.

BROADCAST STATIONS

Country	Frequency (kHz)	Program Time (GMT)	Remarks
CANADA	6160	0930-0500	News each half-hour
	9625 11720	1217-1313	
	9625 11945	2300-2330	
CHINA (People's Rep. of)	7120 9780	0100-0355	Radio Peking
	9480 11685	1200-1255	Radio Peking
	15060	0000-0455	Radio Peking
	17735	0300-0455	Radio Peking
CHINA (Republic of) (Taiwan)	7130 15345 17890	0200-0350	17890 beamed to North America

Country	Frequency (kHz)	Program Time (GMT)	Remarks
	11825 15370 17720	1800-1900	
	3990 7215	0000-2400	American Forces Network Taiwan
CUBA	9525	0630-0800	Radio Havana
EGYPTIAN ARAB REPUBLIC	9475	0200-0330	
GERMAN DEMOCRATIC REPUBLIC (East Germany)	5955 9730	0100-0145, 0230-0315	Beamed to east North America
	5955 6080 6165	0330-0415	Beamed to west North America
GERMAN FEDERAL REPUBLIC	6040 6075 9735	0130-0250	News at 0132
	6075 6145 9545	0435-0555	News at 0437
GREAT BRITAIN	6110	2115-0415	
	9580	2115-0330	
	11780	2115-2315	
	9510	0030-0330	Atlantic Relay (250 kW)
	15260	1500-1515	Saturday only
INDIA	7215 11620	1745-2230	Programs beamed to UK and Western Europe on 312° and 320°. Might receive across North Pole
	9912	1945-2230	
ISRAEL	9009 9625 9725	2045-2130	Programs beamed at Europe
JAPAN	15445 17825	2345-0045	
JORDAN	7155	0955-1315	
	9560	1400-1705	

Country	Frequency (kHz)	Program Time (GMT)	Remarks
LEBANON	11705	1830-2030	May not use English
NORWAY	1578*	0000-0030	Programs in English on Sunday and Monday only, after a 60-minute program in Norwegian
	1578* 9550	0200-0230	
	1578* 6130	0400-0430	*omnidirectional antenna
	6130 9654	0600-0630	
	6130*	1200-1230, 1800-1830	100 or 120 kW transmitting power (except 1578 kHz which is 1.3 or 10 kW)
	17825 21655	1600-1630	
SWEDEN	6175	0030-0100, 0200-0230	
	9630*	1100-1130	*omnidirectional antenna
	11705	0330-0400	
	21505	1400-1430	
UNITED NATIONS	5955 11850	0845-0900	
	15410 21670	1800-1805, 1830-1835	News
UNITED STATES OF AMERICA Voice of America	3980 5965 6160 7270 9635 11915	0300-0730	English programs to Europe
	3980 6040 9760 15205 17785 21455	1600-1800	
	3980 6040 7170 9760 11760	1800-2400	

Country	Frequency (kHz)	Program Time (GMT)	Remarks
American Forces Radio and Television Services	6110 9700 9755 11790 11805 15155 15330 15410 15430 17765 21500	Most times of the day	News, sports, and other timely information to keep US servicemen informed of current events. All broadcasts in English
KGEI (California) (International Broadcast Station)	9760 11880 15280	2230-0500	Various languages Transmitting frequency used depends on month and time of day
WNYW (Radio New York Worldwide)	5985*	0015-0230	*Partly in Spanish
	6075	2230-2345	
	9615*	0130-0230	
	9690	1945-2300	
	9715*	0045-0230	
	11855*	0000-0230	
	11885	2100-2215	
	11890	2000-2100	
	15130*	1830-1945, 2245-0115	
	15215*	2315-0030	
	15440	1900-1945	
	17760*	1700-2400	
	17845	1700-1930 2115-2230	To Europe To Africa
	21525	1700-1815 1700-2100	To Europe To Africa
WINB (World International Broadcasters)	11795	2000-2200	Religious
	17720	1700-2000	Religious
WWV and WWVH	2500 5000 10000 15000 20000*	0000-2400	National Bureau of Standards radio stations at Ft. Collins, Colo. and Maui, Hawaii

Country	Frequency (kHz)	Program Time (GMT)	Remarks
	25000*		Time and frequency standards *Not broadcast by WWVH
USSR (Russia)	41m 31m 25m	2200-2230, 2300-2330 0000-0030, 0330-0730 0030-0100	Radio Moscow (Transmitting frequencies are given by the meter band used.) Radio Tallinn (Tues/Fri/Sun)
	41m 31m	0100-0330, 0400-0530	Radio Moscow
	19m	0330-0500	Radio Moscow
	31m 25m 19m	0320-0330	Radio Yerevan (Sunday)
	11925 15115	1200-1230, 1400-1430.	Radio Tashkent
VIETNAM (Democratic Republic)	7038 10040	2300-2330	Hanoi
	7470	1130-1200, 1430-1500.	Hanoi
	10040 12025	0500-0530, 0830-0900, 1000-1030, 1530-1600, 2000-2030.	Hanoi
	12025 12040	1300-1330	Hanoi
	15018*	1300-1330, 2000-2030.	*15011 kHz is the measured frequency (Hanoi)

SHORTWAVE BANDS

Type	Band	Freq (MHz)	Distance	Best Listening
AIRCRAFT (ssb)	Overseas & military	2.0–9.0	DX	Night
	Overseas & military	9.0–18.0	DX	Day
(vhf/a-m)	Navigation	108.0–118.0	Local/Air	
	Communications	118.0–136.0	Local/Air	

SHORTWAVE BANDS

Type	Band	Freq (MHz)	Distance	Best Listening
AMATEUR RADIO (a-m, ssb, cw)	160m	1.8–2.0	Local	Day/night
	80–75m	3.5–4.0	DX	Night
	40m	7.0–7.3	DX	Night
	20m	14.0–14.3	DX	Day
	15m	21.0–21.45	DX	Day
	10m	28.0–29.7	Short DX	Day
	6m	50.0–54.0	Local	Day/night
	2m	144.0–148.0	Local	Day/night
CITIZENS RADIO (a-m)	Class D	26.965–27.255	Short local	Day/night
INTERNATIONAL SHORTWAVE BROADCAST (a-m)	120m	2.3–2.495	Local	Afternoon
	90m	3.2–3.4	Short DX	Night
	75m	3.9–4.0	Short DX	Night
	60m	4.75–5.06	Short DX	Night
	49m	5.95–6.2	DX	Sunset, night
	41m	7.1–7.3	DX	Sunset, night
	31m	9.5–9.775	DX	Day/night
	25m	11.7–11.975	DX	Day/night
	19m	15.1–15.45	DX	Day
	16m	17.7–17.9	DX	Day
	13m	21.45–21.75	DX	Day
	11m	25.6–26.1	Short DX	Day
MARINE (ssb)	Coastwise	2.0–2.5	Local	Day/night
	High seas & inland waterways	4.0–9.0	DX	Night
	High seas	12.0–22.0	DX	Day
(vhf-fm)	Inland waterways	156.0–162.0	Local	Day/night
NATIONAL WEATHER SERVICE (vhf-fm)		162.4, 162.45, 162.55	Local	Day/night
SAFETY AND COMMERCIAL (fm)	Low-band vhf	30.0–50.0	Local	
	High-band vhf	152.0–174.0	Local	
	Uhf	450.0–470.0	Short local	